AGRICULTURAL POLICY REFORMS AND REGIONAL MARKET INTEGRATION IN MALAWI, ZAMBIA, AND ZIMBABWE

AGRICULTURAL POLICY REFORMS AND REGIONAL MARKET INTEGRATION IN MALAWI, ZAMBIA, AND ZIMBABWE

Edited by Alberto Valdés and
Kay Muir-Leresche

International Food Policy Research Institute (IFPRI)

Washington, D.C.

Published in 1993 by the

International Food Policy Research Institute
1200 Seventeenth Street, N.W.
Washington, D.C. 20036
U.S.A.

Library of Congress Cataloging-in-Publication Data

Agricultural policy reforms and regional market integration in Malawi, Zambia, and Zimbabwe / edited by Alberto Valdés and Kay Muir-Leresche.

p. cm.

Includes bibliographical references.

ISBN 0-89629-327-0

1. Agriculture and state--Africa, Southern. 2. Grain trade--Government policy--Africa, Southern. 3. Africa, Southern--Economic policy. 4. Africa, Southern--Commercial policy. 5. Africa, Southern--Economic integration. 6. Southern African Development Coordination Conference. I. Valdés, Alberto, 1935- . II. Muir-Leresche, Kay.

HD2130.Z8A37 1993 93-18029
338.1'868--dc20 CIP

Contents

Tables and Figures

Tables

Figures

Foreword

Economic growth in developing countries has generally slowed since the early 1980s. In many countries in Sub-Saharan Africa, per capita income has even declined. External developments have had an adverse effect, as the slowdown in world economic activity and deteriorating terms of trade for most agricultural commodities have posed severe difficulties for many developing countries. More important, however, inappropriate domestic policies have been increasingly recognized as a major culprit. This has led to the adoption of various programs of macroeconomic and sectoral policy reforms, generally aimed at achieving macroeconomic stability and a satisfactory rate of economic growth. Among southern African countries, there has also been keen interest in promoting regional market integration.

Given the predominance of agriculture in low-income economies, it is not surprising that poor performance of the agricultural sector has been associated with instability in the macroeconomy and unsatisfactory growth of national income. One can reasonably expect also that the promotion of agriculture's interest will be critical to the success of economic policy reforms in those countries.

Getting the policies right for agriculture is likely to require reforms not only of policies directed specifically to the agricultural sector but also of government interventions elsewhere in the economy that indirectly affect agricultural incentives and performance. Indeed, a growing body of evidence strongly indicates that the incentive bias against agricultural producers arising from trade and macroeconomic policies has tended to outweigh the direct effect of agricultural sector-specific policies.

IFPRI research has long recognized the need to take account of the influence of trade and macroeconomic policies on relative incentives and constraints facing agricultural producers. This has been the subject of several country studies—on Argentina, Chile, Colombia, Nigeria, Pakistan, Peru, the Philippines, and Zaire—undertaken since the early 1980s, resulting in a number of contributions to the IFPRI Research Report series and outside publications. More recently, the Trade and Macroeconomics Division has begun work on developing-country experiences in the formulation, implementation, and impact of economic policy reforms affecting food and agriculture.

The possibilities not only of multilateral trade expansion but also of increased intraregional trade among developing countries have been analyzed by IFPRI researchers. In particular, expansion of regional food markets is viewed in these studies as contributing to the reduction of commodity price variability in the cooperating countries and providing more efficiently for the distribution of commercial inputs to agricultural production. Given the renewed interest in the intensification of regional cooperation efforts, regional trade arrangements to promote food security and agricultural trade will continue to be in the research agenda of the Trade and Macroeconomics Division for at least the next few years.

The present volume addresses some of the important issues concerning economic policy reforms as they relate to the food and agricultural sector and regional integration in agricultural markets among developing countries. It is especially significant that the interface between these two areas of policy concern is explicitly examined and that "agricultural policy" is viewed in a macroeconomic perspective. The empirical focus is on three countries in southern Africa, but the analytical approaches and policy implications are of wider applicability, being relevant to many other low-income countries.

Romeo M. Bautista
Acting Director
Trade and Macroeconomics Division

Acknowledgments

This volume is a product of a three-year project on regional integration of agricultural markets in southern Africa, which is a component of a wider study supported by a research grant to IFPRI by the Swiss Development Cooperation and Humanitarian Aid. The Swiss grant financed not only the research and writing of this volume but also the two policy workshops held in Zimbabwe in February 1990 and November 1991 that IFPRI cosponsored with the Ministry of Lands, Agriculture and Rural Resettlement, the Food Security Unit of the Southern African Development Coordination Conference, and the University of Zimbabwe.

In these workshops the participation of senior policymakers and analysts in the region provided a critical perspective and insightful comments on the papers presented, which have been revised for inclusion in this volume. The participation of Dominic Mulaisho, then the editor in chief of *Southern African Economist* and now governor of the Central Bank of Zambia, in the second workshop was particularly valuable. Comments on individual chapters were made by Charles Mann, Joseph Mhango, J. B. Wyckoff, Doris Jansen, and Jim Shaffer. Dean DeRosa of IFPRI and William Masters of Purdue University read through an earlier draft of the entire volume and gave written comments and suggestions for revision as part of the review process for this IFPRI publication.

We would like to extend our appreciation to Marcelle Thomas for her research and editorial support to the editors and authors and for her role as a key coordinator throughout the duration of this project, and to Susan Frost, who has provided extensive secretarial and organizational support.

We are grateful to these institutions and individuals for their contributions to the success of the workshops and the preparation of this book.

List of Contributors

Roger W. Buckland is a technical advisor in the Food Security Technical and Administrative Unit of the Southern African Development Coordination Conference, Harare, Zimbabwe.

Thomas S. Jayne is a visiting assistant professor at Michigan State University. He was formerly a visiting lecturer at the University of Zimbabwe.

Share J. Jiriyengwa is an economist and head of the Planning Unit at the Grain Marketing Board in Harare, Zimbabwe.

Ulrich Koester is a professor of agricultural economics and director of the Institut fuer Agrarpolitik und Marktlehre, University of Kiel, Germany.

Katundu M. Mtawali is principal economist in the Ministry of Agriculture in Lilongwe, Malawi.

Kay Muir-Leresche is a senior lecturer in the Department of Agricultural Economics and Extension at the University of Zimbabwe.

Ernst-August Nuppenau is a lecturer in the Department of Agricultural Economics at the University of Kiel, Germany.

Thomas C. Pinckney is an assistant professor in the Department of Economics at Williams College, Massachusetts. He was formerly a research associate at the International Food Policy Research Institute and before that in the Department of Economics at the University of Nairobi and in the Economic Research Bureau at the University of Dar-es-Salaam.

Jorge Quiroz is a professor of economics at the Instituto Latino Americano de Desarollo Económico y Social in Santiago (ILADES), Chile.

Julius Shawa is acting principal economist in the Planning Division in the Ministry of Agriculture, Food, and Fisheries in Lusaka, Zambia.

Tobias Takavarasha is deputy secretary for economics and marketing in the Ministry of Lands, Agriculture and Rural Resettlement in Harare, Zimbabwe.

Alberto Valdés is principal agricultural economist for Latin America at the World Bank. He served as director of the International Trade and Food Security Program at the International Food Policy Research Institute from 1976-1990.

1
Introduction and Overview

Kay Muir-Leresche and Alberto Valdés

The 1990s promise to be a decade of change for southern Africa, as for most of the world. These changes have been precipitated by both political and economic forces. The major economic incentives for change are the continuing decline in per capita income for many countries and the static or declining health and education facilities. Immediately after independence, countries such as Tanzania embarked on aggressive social programs that saw a marked improvement in social indicators. These advances were slowed and even reversed in later decades, as the economic growth to sustain these activities declined. The political changes in Eastern Europe, combined with increasing evidence of the importance of encouraging competition as a necessary condition of economic growth, have resulted in pressures for the heavily centralized economies in southern Africa to liberalize.

Most of the chapters presented in this volume concern the current status of and potential for liberalization of foodgrain trade in Malawi, Zambia, and Zimbabwe; two chapters—one by Koester and the other by Quiroz and Valdés—include Tanzania in the analysis. The research is concerned with the overall economic environment and with its impact on household economies. There has been much pressure from international funding agencies for the economies in the region to reform. Reforms implemented during the 1980s in some of the countries have been seriously undermined by the partial nature of the reforms. The reforms in Zambia were accepted (but seldom fully implemented) only until external donor funds were received, by which time popular opposition to the rising food costs that resulted from first-round adjustments were used by the government in power as an excuse to abandon the reforms despite undertakings to international agencies.

A similar scenario could be enacted in Zimbabwe, which appears, 18 months after the announcement of a structural adjustment program, to have made very little effort to reduce government expenditure and liberalize its domestic markets or trade, although it has attempted to reduce food subsidies and introduce reforms of its foreign-exchange policy. It has also made some effort to hold down expenditure by freezing the number of posts in the bureaucracy but has made no attempt to reduce its plethora of ministries and the size of its cabinet. There has been much discussion of attempts to rationalize the Grain Marketing Board, but no positive steps have been taken to decontrol the other marketing boards—for example, the Dairy Marketing Board and Cold Storage Commission—which would have few, if any, negative impacts on rural development or the poor. The politics of power still appear to override economic reality, and there has been little deregulation in an economy where barriers to entry abound. For example, it is illegal for all but a few firms to mill and sell the staple maize meal in urban areas, and it is illegal to subdivide land, so the country is locked into large-scale production on half of its land area and increasing subdivision of overpopulated communally owned land on the other half.

Malawi's adjustment program has made some progress toward increasing competition and has recently made efforts to cushion the poor, but the historical bias toward the large-scale sector, particularly since independence, still tends to marginalize peasant farmers despite deliberate attempts to reduce implicit taxation of the sector. Malawi was almost totally undeveloped at independence and had to borrow heavily to develop its infrastructure and assist with recurrent expenditure. Its efforts to improve social services are severely constrained by its foreign-debt servicing, poor geographical position, and limited resources.

The most important and revealing aspect of this document is its discussion of the still-limited liberalization of agricultural marketing in southern African countries, compounded by deep-rooted concerns among policymakers for food security. Thus, barriers to internal trade in agriculture remain high and, as emphasized by Koester (Chapter 3), liberalization of external trade should proceed hand in hand with liberalization of internal trade. In addition, although Valdés (Chapter 2) contends that the debate over agricultural trade and price policies is no longer over economic fundamentals, it is notable that many of the chapters contributed by regional officials indicate only a limited recognition and acceptance of the broader tenets of liberal economic theory. While policymakers in the region seem to accept the need for important changes in agricultural-sector policies, their commitment to changes in broader institutional and industry-protection policies that would have equally large, if not larger, direct and indirect effects on agricultural output and trade is problematic. Responsibility for this lies in part with the bureaucracy and others who have vested interests in maintaining existing food-security programs, agricultural parastatals, and more broadly administered agricultural and trade policies. But more fundamentally, the political economy of fostering broader economic reforms in Africa may still be bounded by unfortunate social and political conditions, including underdeveloped communications facilities and human resources, that impede the pace of political as well as economic reforms in the region.

Concentrating on the situation in southern Africa, these studies use the maize markets in Malawi, Zambia, and Zimbabwe to consider the rationale for trade liberalization and its effects on development and food security. The chapters are divided into four broad but interrelated topic areas: the general policy environment for liberalization; national policy and liberalization; the effects of internal and market reforms on food security; and the potential for expanded intraregional trade.

GENERAL POLICY ENVIRONMENT FOR LIBERALIZATION

The chapter by Valdés highlights both the necessary conditions and the need for countries to move toward less-regulated, more-open economies. It includes a review of empirical findings that indicate that macroeconomic policies have had a major impact and have resulted in large income transfers from agriculture to government and nonagriculture. These arise principally from the indirect effects of industrial protection and exchange-rate misalignment. The chapter also provides a potentially useful set of guidelines for identifying and appropriately sequencing needed trade, macroeconomic, and sectoral policy reforms; specifically, it emphasizes the need to maintain an increase in the real exchange rate to continue to stimulate exports and other tradables throughout the program, and the urgent need to dismantle quantitative trade and marketing restrictions and ensure transparency.

Koester briefly demonstrates that trade among the countries could develop even if there were no internal changes, but also shows how synchronized internal decontrols would reinforce the benefits from trade. He summarizes the main arguments for liberalization and greater regional market integration in the specific context of Malawi, Tanzania, Zambia, and Zimbabwe. Koester shows how the burden of liberalizing internally could be reduced if countries were to synchronize these efforts and open their borders.

Quiroz and Valdés extend the Schiff-Valdés approach of the direct and indirect effects by decomposition of the changes in agricultural relative domestic prices into the determinants of these changes, namely, border prices, real exchange rate, and trade policy. This approach is applied to Tanzania as well as Malawi, Zambia, and Zimbabwe.

NATIONAL POLICY AND LIBERALIZATION

Takavarasha reviews the need for reform in Zimbabwe. He shows that the economy was highly regulated and controlled at independence and that this has continued, although the strongest and wealthiest have often been the beneficiaries of the situation. Increasing transparency and access to market information are essential to highlight past anomalies and to open up opportunities for wider participation. Too many years of controls and the continuation of unrecognized regulatory barriers are a serious constraint to the liberalization and structural-adjustment process.

Shawa gives an overview of Zambia's experience with reform programs. He shows that Zambia's past failures were primarily the result of a lack of serious political commitment. The new government appears to be more serious about reform, and Shawa considers that there is less likelihood that history will repeat itself.

Mtawali's overview of the situation in Malawi shows that the reforms have been partial, and while efforts have been made to contain government spending, this has been at the expense of social services. The reforms are undermined by foreign-debt interest and principal repayments. Mtawali shows that the grain-market reforms have been successful in reducing parastatal deficits and increasing efficiency and producer prices in accessible markets, but they have reduced market access in remote areas.

The reforms described in these three chapters hold considerable promise that the extent of unnecessary government intervention in agriculture will be reduced in the future, particularly by fostering a return to the primacy of traders and other private-sector agents operating in predominantly unregulated markets for food and other agricultural commodities. But a skeptical view is that a firm commitment to thoroughgoing liberalization of the agricultural sector, accomplished through overall as well as sectoral policy reforms, is still lacking on the part of most policymakers in Africa. Indeed, the slow pace of the reforms to date and the primary concern for food security through self-reliance and continued food aid belies the limited acceptance of the liberal economist's prescription for greater growth and economic welfare through broad liberalization of domestic markets and external trade relations.

EFFECTS OF INTERNAL AND MARKET REFORMS ON FOOD SECURITY

Pinckney shows that there are legitimate concerns that cause governments to intervene in grain markets, particularly where there is high production instability and where external supplies are uncertain or of a less-desired quality (white versus yellow maize) or both. He goes on to consider the domestic market intervention options and suggests that a policy that dampens price movements would be less economically inefficient than a set price band. He acknowledges, however, that the dampening system would be more open to abuse than floor and ceiling bands and, furthermore, that it may require more sophisticated market information and price analysis expertise than currently exists in these countries. Pinckney recommends that government goals on stockholding be clearly articulated and the trade-offs made explicit and all distortions transparent.[1]

Jayne and Nuppenau consider the impact of various levels of domestic market liberalization of maize in Zimbabwe. They develop a framework that is useful both as an empirical analysis of current policy and as a prediction of the directions of the effects of the changes on producers and consumers—the actual results depend upon various assumed relationships so that their magnitude may be questioned, but the direction is clear. This framework is useful in highlighting the trade-offs and can be developed for agricultural policy analysis units in other countries in the region. The chapter highlights the incongruity of spending enormous resources investigating how to make parastatals more efficient, when a few simple regulations could be removed, improving the welfare of the treasury, consumers, and producers and reducing only that of the milling oligopolies.

Jiriyengwa highlights the problems arising from associating the structural adjustment program in Zimbabwe with budgetary cost-saving rather than economic rationalization. He also considers the social and political implications of a reform program that does not recognize the

---

[1] Important empirical and theoretical studies suggest that price instability might be offset in part by risk-spreading investment and output, and that at least macroeconomic performance is not clearly adversely affected by the instability of commodity prices (Behrman 1987). Given the questionable economic record of parastatal marketing boards and strategic grain reserves in Africa, it is difficult to endorse at face value the view presented in the paper that government agencies might be counted upon to stabilize prices more efficiently in the future, following modest "leaning against the wind" or other enlightened operational rules generated by econometric models.

nutritional and developmental role of maize for both the urban and rural poor. He gives a detailed breakdown of the financing of the various activities of the Grain Marketing Board and suggests retaining the board, which would compete in the local market, but that its services to remote areas and its stabilization functions should be directly subsidized by the government. This raises the issue of transparency of the costs of agricultural-marketing boards and food-security agencies. Also, to the extent that the origins of these organizations can be traced in part to similar arguments for their expansion to wider and wider commercial activities in earlier years, the argument for continued commercial operations by such government entities, crowding out private-sector traders and other intermediaries, is an open question.

Buckland gives an overview of the food-security program of the Southern African Development Coordination Conference (SADCC) and the expanded role of the regional body with national liberalization. Conceptually, trade liberalization is a substitute for commodity stockpiling. Thus, the food-security unit of SADCC must change its focus in a new trade-policy environment. The proposals offered by Buckland emphasize recasting the food-security program to entail greater provision of agricultural extension services, including information and communications systems, and incorporation of policy components concerned with the environment and sustainable agricultural development in southern Africa. Finally, Buckland proposes that the program incorporate greater analytical capacity to study and evaluate the implications of trade liberalization for food security, especially among disadvantaged groups.

Nuppenau explores the effect of opening up maize trade between Malawi, Zambia, and Zimbabwe without liberalizing internally. His model assumes that the same price relationships would exist after trade as existed before. The model is in fact a transport model that shows how freer trade would reduce transport costs and increase welfare even if the current distortions existed. The paper, therefore, shows Zambia as a net importer of maize, which reflects current reality in the light of the internal distortions. The agronomic potential is, in fact, for Zambia to be the major maize supplier in the region, but this would involve rationalization and liberalization of the markets within the countries.

POTENTIAL FOR INTRAREGIONAL TRADE

Koester explores in more depth the specific policies required to promote trade between the countries. He discusses the role of government and presents some guidelines for phasing in agricultural trade promotion policies.

Muir-Leresche considers the impact of a free South Africa on trade and resource allocation in the region. She shows that the internal reforms and rationalization of the macroeconomic policies within the southern African countries are likely to have a much greater impact on increasing agricultural trade than closer relations with South Africa without reform. The full advantages of closer ties with South Africa could only be realized if all the countries of the region, including South Africa, encouraged prices that reflect equilibrium in both the capital, goods, and service markets. She considers the potential for successful reform within the region and highlights the importance of ensuring that the reforms are made within an environment that encourages competition. Politicians need to be made aware of how the reform programs could be completely undermined by unrecognized barriers to entry, so that they are more prepared to take the risks of passing control to the market.

REFERENCE

Behrman, J. R. 1987. Commodity price instability and economic goal attainment in developing countries. *World Development* 15 (May): 559-573.

government and presents some guidelines for designing agricultural trade promotion policies.

[illegible] considers the impact of a free South African trade and [illegible] on the region. She states that the internal reforms and liberalization [illegible] of the macroeconomic [illegible] within the southern African countries are likely to have a much greater impact on intraregional agricultural trade than closer relations with South Africa without reform. The full advantages of [illegible] with South Africa could only be realized [illegible] of the [illegible] South [illegible] goods, and service markets. She [illegible] the potential for [illegible] within the region and highlights the importance of [illegible] that the reforms [illegible] within [illegible] that [illegible] of how the reform [illegible] completely understood [illegible] and barriers to entry [illegible] they are more prepared [illegible] the risk of [illegible] control to the market.

REFERENCES

[illegible] J. R. 1987. Commodity price instability [illegible] in developing countries. World Development 15 (May): [illegible]

Part I
General Policy Environment

2
The Macroeconomic and Overall Policy Environment Necessary to Complement Agricultural Trade and Price Policy Reforms

Alberto Valdés

The debate on agricultural trade and price policy is no longer about the fundamentals. It is no longer about the basic course of economic policy that most developing countries want to follow, but rather the problems of transition toward the new policies.

In many developing countries there is a new attitude toward economic policies and the role of the public sector. The changing economic and political realities in so many developing countries call for a new development strategy, and this seems to be an unprecedented opportunity for their agriculture. Many countries are embarking on a revision of trade and pricing policies, moving toward a more open economy, and recognizing the importance of maintaining fiscal discipline and a realistic exchange rate to achieve a broad-based, sustainable, economic growth.

Agricultural policy reform is a particularly complex process. Historically, the agricultural sector has been subject to extensive government interventions in both output and factor markets. It is a sector whose structure of costs and returns depends substantially on the performance of other sectors, such as transport and communications, and related markets, such as the financial and foreign-exchange markets. These related sectors are in turn subject to extensive interventions. Further, the sector comprises thousands of small productive units that independently make day-to-day decisions about on-farm investments and cropping patterns and on whom the credibility and cohesiveness of the policy-reform package will have a profound influence.

One complex question for policymakers is how broad an effective reform process must be and what sequence the reform measures should follow. Studies on the lessons from policy-reform experiences in other countries indicate a possible strong interaction between the macroeconomic process and the response to reforms by the agricultural sector.

For example, to the extent that Zimbabwean farmers suffer from restrictions on access to production inputs and that regulations prevent competition in the input delivery system, the output response from output market reforms would be greatly constrained. Furthermore, delineating the boundaries for the microeconomic aspects of liberalization raises complex issues that remain to be fully analyzed.

What should be the role of state agencies in agricultural marketing during and after the trade reform? How to deal with a variety of sectors currently subjected to extensive regulations—such as the financial, labor, and land markets and the transport and communications systems—is critical, because bottlenecks in these related sectors could inhibit the output response as well as the credibility and sustainability of agricultural trade and price reforms.

The linkages between agricultural trade and price reform and the deregulation of internal markets are critical in the reform process. Implementing the former without internal market deregulation is unlikely to have much effect on investment and productivity. Internal regulations often prevent entry and competition in the domestic input and output markets and impose serious barriers to the success of policy-reform efforts in Sub-Saharan Africa.

The following section presents a synthesis of the findings in the study directed by Schiff and Valdés (1992) for 18 developing countries for the 1960-84 period. The next two sections offer evidence on the effect of price interventions in the past, which supports the argument for policy reform. Evidence is included on the patterns of agricultural protection and on the effect of this intervention on farm income (other effects are presented in Schiff and Valdés 1992). The last section identifies critical elements for a new agricultural trade strategy. The Appendix sets up a basic concept of an efficient structure of incentives from an economywide perspective against which the actual incentives policies can be evaluated.

WHAT HAVE WE LEARNED?

Based on the analysis by Schiff and Valdés (1992) for 18 countries in Asia, Sub-Saharan Africa, North Africa and the Middle East, and Latin America during 1960-84, the major findings can be summarized around the following nine conclusions.

1. Industrial protection and macroeconomic policies are, in the long term, the principal determinants of the structure of incentives to agricultural production. Approximately two-thirds of the total effect of price interven-

tions on incentives comes from the effect of economywide policies (indirect effect), while the remaining third comes from the effect of price interventions specific to the agricultural sector (direct effect).[1] Using the nominal rate of protection as a measure for these effects, it was found that on average for these 18 countries throughout the 25 years examined, the indirect price interventions were equivalent to a tax of 22 percent—approximately three times the effect of agricultural (direct) policies that resulted in a tax of 8 percent. The total level of price interventions (that is, direct plus indirect) was minus 30 percent.

2. During the 25 years analyzed, industrial protection was more influential in most countries than the exchange-rate misalignment on agricultural incentives. More recently, however, the relative weight of these two factors seems to be changing toward an increasing influence of macroeconomic variables.

3. The high taxation of agriculture from the direct and indirect price interventions—to a large extent an implicit tax—was statistically associated with a lower agricultural and overall growth rate. Parametric and nonparametric tests were performed to test this association.

4. Historically, most developing countries have protected their agricultural import-competing subsector, largely dominated by food-crop production. Using the official exchange rate, the computed average (equivalent) nominal rate of protection to this subsector during 1960-84 was approximately 18 percent.

5. In contrast, most countries taxed the production of agricultural exportables. At the official exchange rate, the average nominal rate of protection was minus 16 percent. That is, within agriculture, a systematic pattern of direct price interventions protecting importables and taxing exportables is observed. The indirect interventions reinforced the taxation to exportables and substantially reduced the net protection to importables.

6. Direct price interventions reduced domestic price instability relative to that which would have prevailed under no interventions. Agricultural price interventions reduced domestic price instability by approximately 25 percent on average. Indirect price interventions did not contribute to domestic price stability.

7. As a result of the combined effect of direct and indirect price interventions, there was a huge transfer of income from agriculture to the government and the rest of the nonagricultural sectors. These income transfers represented more than 30 percent of agricultural gross domestic

[1] See the Appendix for details on the definition of direct and indirect effects.

product (GDP) during the 25 years and were substantially higher for the three African countries (Côte d'Ivoire, Ghana, and Zambia) than for the rest of the countries studied. The transfers to agriculture via input and credit subsidies, and through government investment, represented a very small fraction of the transfer out of agriculture.

8. There is no evidence to indicate that the removal of the direct and indirect price interventions would have been regressive. Only estimates of the short- and medium-term effects on income distribution are available because it is practically impossible to measure ex post the longer-term effect of price interventions on income distribution. Removal of direct and indirect interventions would affect employment, rural wages, and rural-urban migration in addition to gross and net farm income. Based on the estimates available from this study, it was found that with few exceptions (Egypt, Argentina) the short-term real-income effect of the direct price interventions on household income was quite small (positive or negative), in most cases lower than 5 percent of household income. In the long run, through its effect on rural employment and farm income, the removal of price interventions probably would have favored the rural poor in the countries studied. It is unrealistic to expect that the policy analyst can measure ex ante the longer-term effects of the policies on income distribution. Thus, price interventions to achieve social objectives are not based on firm empirical evidence. A policy matrix approach and computer general equilibrium (CGE) models would not do the job for the long-run analysis.

9. Fiscal revenues have been an influential motivation behind the explicit taxation of exportables in some countries. However, the importance of these revenues (relative to total public revenues) has been declining steadily in almost all the countries studied. In a few countries, for instance, Zambia, Sri Lanka (before the late 1970s) and Egypt, food and input subsidies were significant sources of government expenditures and in most cases were unsustainable with domestic resources.

These results have important implications for the design of trade and price policies for agriculture, which is discussed in the last section of this chapter.

PATTERNS OF AGRICULTURAL PROTECTION

A most striking result shown in Table 2.1 for 16 of the countries studied by Schiff and Valdés (1992) is the high level of price interventions in agriculture. For all products and all years, total price interventions (*NPRt*) averaged approximately −30 percent. In other words,

Table 2.1—Average direct, indirect, and total nominal protection rates

Region/Country	Period	Indirect	Direct Importable	Direct Exportable	Direct Protection	Total Protection
			(percent)			
Sub-Saharan Africa		-28.6	17.6	-20.5	-23.0	-51.6
Côte d'Ivoire	1960-82	-23.3	26.2	-28.7	-25.7	-49.0
Ghana	1958-76	-32.6	42.9	-29.8	-26.9	-59.5
Zambia	1966-84	-29.9	-16.4	-3.1	-16.3	-46.2
North Africa		-18.5	-6.7	-25.7	-19.9	-38.4
Egypt	1964-84	-19.6	-5.1	-32.8	-24.8	-44.4
Morocco	1963-84	-17.4	-8.2	-18.5	-15.0	-32.4
Africa total		-24.6	7.9	-22.6	-21.8	-46.3
East Asia						
Korea	1960-84	-25.8	39.0	n.a.	39.0	13.2
South Asia		-32.1	16.1	-12.0	-7.7	-39.8
Pakistan	1960-86	-33.1	-6.9	-5.6	-6.4	-39.5
Sri Lanka	1960-85	-31.1	39.0	-18.4	-9.0	-40.1
Southeast Asia		-15.5	20.5	-16.3	-12.9	-28.4
Malaysia	1960-83	-8.2	23.6	-12.7	-9.4	-17.6
Philippines	1960-86	-23.3	17.4	-11.2	-4.1	-27.4
Thailand	1962-84	-15.0	n.a.	-25.1	-25.1	-40.1
Asia total		-22.8	22.4	-14.6	-2.5	-25.3
Latin America						
Argentina	1960-84	-21.3	n.a.	-17.8	-17.8	-39.1
Brazil	1969-83	-18.4	20.2	5.4	10.1	-8.3
Chile	1960-83	-20.4	-1.2	13.5	-1.2	-21.6
Colombia	1960-83	-25.2	14.5	-8.5	-4.8	-30.0
Dominican Republic	1966-85	-21.3	19.0	-24.8	-18.6	-39.9
Latin American total		-21.3	13.1	-6.4	-6.5	-27.8
All-region average		-22.5	14.4	-12.6	-7.9	-30.3

Source: Based on M. Schiff and A. Valdés, *A Synthesis of the Economics in Developing Countries*, vol. 4 of *The Political Economy of Agricultural Pricing Policy* (Baltimore, Md., U.S.A.: Johns Hopkins University Press, 1992).

Note: n.a. means not available.

in the absence of price interventions, the relative price of agricultural goods would have been 42 percent higher (30/70). Important differences among countries emerged. While Côte d'Ivoire, Ghana, and Zambia exhibited the highest degree of negative price intervention (*NPRt* equal to -51.6 percent on average), the degree of price intervention for Malaysia and Brazil, for example, was substantially lower (*NPRt* between -18 percent and -8 percent) although still negative, and in Korea, agriculture was protected (*NPRt* of 13 percent).

A second important result concerns the source of price interventions. On average, the indirect price interventions accounted for approximately three-fourths of the total disprotection toward agriculture. This high negative indirect intervention arose mainly as a result of the high prevailing levels of industrial protection and, to a lesser although still influential extent, from the exchange-rate misalignment resulting from both macroeconomic imbalances and industrial protection.

An important finding concerning the direct price interventions is the systematic difference observed in the treatment of importables vis-à-vis exportables, resulting in a strong antitrade bias. While direct price interventions to agricultural import-competing activities were in most cases positive (*NPRd* between 7.9 percent and 22.4 percent for the country groups in Table 2.1), direct protection to agricultural exportables was in most cases negative (between -6.4 percent and -22.6 percent). On average for the 16 countries, direct intervention resulted in a protection rate of about 14 percent for importables and a tax rate of nearly 13 percent for exportables.

This pattern of direct protection is attributed to the desire to achieve a certain minimum level of self-sufficiency in food production in the case of importables, and to collect government revenues in the case of exportables.[2] Supporters of agricultural protection in Japan, Sweden, and the European Economic Community (EEC) have made persuasive use of the food-security objective as an argument for protection of food products (Honma and Hayami 1986). This is often justified by the gloomy picture of world demand and supply for cereals and the risks of food shortages. The findings for this sample of less-developed countries (LDCs) suggest that a relatively high weight was also given in these countries to this food-security objective, regardless

---

[2] For the sample of the 18 countries, it was estimated that the budget effect of direct price interventions on agricultural exportables contributed to approximately 20 percent of total public expenditures during 1960-69, 11 percent during 1970-79, and 5.8 percent during 1980-83 (Schiff and Valdés 1992).

of whether the relevant world food market in question was very thin (such as white maize) or was a fairly well developed market (such as wheat) with central transaction points.

These findings suggest that there was a substantial resource misallocation between importables and exportables. The optimal export tax argument cannot be used as a defense for this taxation of exportables except in a few cases. Preliminary findings of a recent study by Panagariya and Schiff (1990) indicate that for 1986 the optimal export tax on cocoa in Côte d'Ivoire was 25 percent and in Ghana about 20 percent; hence the level of their export taxes was not too far off from the optimum tax (however, rice was highly protected, so P_m/P_x is still distorted).[3] For cotton in Egypt, the actual direct export tax (32 percent) was below the optimum (53 percent). Coffee in Brazil and Colombia, rice in Thailand, and rubber in Malaysia are other relevant cases. Similarly, Zambia has market power in white maize, hence the marginal import cost is higher than the border price, and an import tariff should have been imposed; however, the present study indicates an average direct tax on importables of 16 percent. Thus, allowing for optimal trade interventions for nonpricetakers reduces the degree of the actual distortion (from direct price interventions) to exportables in a few cases. But the case of nonpricetakers is not the rule, and when it does apply the actual tax levels often do not coincide with the optimum levels. Relative prices within agriculture are distorted even when adjusted for the optimal trade tax.

Undoubtedly, there is widespread concern that policy reform will adversely affect the poor.

One of the striking results of the comparative study of the 18 LDCs by Schiff and Valdés (1992) is how little the direct price interventions have, historically, achieved in alleviating poverty. Often they benefited the urban poor at the expense of the rural poor rather than the urban rich. The bulk of the income transfer (in the short to medium term) resulted from the economywide policies and not from the sectoral price interventions. In most LDCs, farmers received only a fraction—about 40 percent—of what they would have received under no intervention. That is, farmers' income would have been 2.5 times higher.

This short- to medium-term impact of price intervention on real income of households can be measured. However, it is virtually impossible to anticipate the longer-term effects of a broad policy reform

---

[3] P_m/P_x expresses the price of importables relative to the price of exportables.

on income distribution. Changes in rural and urban demand for labor and effects on rural-urban migration, changes in market prices of household purchases, and changes in the output mix and aggregate land of production are variables that are extremely hard to estimate as would be required for a meaningful empirical study of their impact on poverty.

AGRICULTURAL TRADE AND PRICE POLICY REFORM: IMPLICATIONS FOR FARMERS' INCOME

Price interventions can affect agricultural growth, consumption, and trade flows. But agricultural price interventions also have other, broader economic implications through their influence on the government budget and on the real income of urban and rural households. As with most policy interventions, in the process there are winners and losers, an issue on which more empirical research is needed for a better understanding of the motivations underlying price interventions and the political and economic constraints to policy reform.

In this section, the focus is on one dimension, namely, the consequences for farm income. The background material that provides the actual estimation of these effects is found in Schiff and Valdés 1992.

Price interventions can generate substantial resource transfers within and between sectors. In fact, one motivation for the policies is precisely the inducement of these transfers. Transfers from direct price interventions may take a variety of forms: export taxes that provide a source of government revenue; food subsidies through lower food prices to urban consumers; input subsidies such as fertilizers and credit; parastatals that capture the revenues from trade by their monopolistic status on agricultural trade but also suffer losses from selling at lower domestic prices. Transfers from indirect price interventions will result from the exchange-rate misalignment and the effect of industrial protection on the prices paid by farmers for product inputs and consumer goods.

Price-related transfers were defined as the change in real income of agriculture that result from direct and indirect price interventions affecting output and input prices and the prices of consumer goods purchased by farm households (Schiff and Valdés 1992). Specifically, these transfers were measured as the change in value added resulting from both price interventions measured at the actual level of production and adjusted for the change in the rural consumer price index.

Nonprice transfers were defined to include public investments that can be considered public goods, such as irrigation, roads, research, and extension. Marketing-related expenditures by state agencies, such as storage, were excluded on the premise that (1) these interventions are reflected in domestic prices paid or received by farmers and thus appear under price interventions, and (2) they do not clearly constitute a transfer to agriculture, considering that these activities could be, and in many countries are, undertaken by private traders.

As an illustration of the magnitudes of income transfers, the results on income transfers from total price interventions for three Sub-Saharan African countries during 1960-84 are presented in Table 2.2, all expressed as a percentage of agricultural GDP. The results show that total price interventions on outputs reduced agricultural GDP by about 28 percent, and transfers into agriculture through input subsidies raised agricultural income by approximately 8 percent. Expanding the output coverage to the rest of agriculture (given that input subsidies apply to most of agriculture) and assuming that there are no direct price interventions for the rest of agriculture (that is, nominal rate of protection for the rest equals zero) raises the average net transfer out of agriculture as a result of total price interventions to a staggering 103 percent. Finally, the non-price-related transfers into agriculture amounted to approximately 8 percent, resulting in a net overall transfer out of agriculture of approximately 96 percent of agricultural GDP.

While input subsidies and public investment do to some extent "compensate" for the negative transfer through output prices, albeit in an inefficient form, as shown in Table 2.2 this compensation is equivalent to only a fraction of the income loss, particularly when the indirect effects are taken into account.

The magnitude of these net transfers out of agriculture is astonishing. Its cumulative effect through time must have been profoundly harmful to the sector in the long run in terms of lowering farm investment and income.

SOME GUIDELINES FOR A NEW AGRICULTURAL TRADE STRATEGY

Recognition is growing that governments are burdened with economic functions that they are incapable of performing efficiently, while many government functions that cannot be provided by the private sector are neglected, such as primary education in rural areas, management of land titles, construction of roads, and agricultural research.

Table 2.2—Net income transfers to and from agriculture as a result of direct and indirect price and nonprice interventions, 1960-84

		Total Price Transfers							Sum of Total Price and Nonprice Transfers		
				Output of Other Agricultural Products		Sum of Total Price Transfers					
Country	Period	Output of Selected Products (1)[a]	All Inputs (2)[b]	Assumption 1 (3)[c]	Assumption 2 (4)[c]	Assumption 1 (Sum of Columns 1, 2, & 3) (5)	Assumption 2 (Sum of Columns 1, 2, & 4) (6)	Nonprice Transfers (7)[d]	Assumption 1 (Sum of Columns 5 & 7) (8)	Assumption 2 (Sum of Columns 6 & 7) (9)	Average of Assumptions 1 & 2 (Half of Columns 8 & 9) (10)
		(percent of agricultural GDP)									
Côte d'Ivoire	1960-69	-13	1	-10	-55	-22	-67	6	-16	-61	-38.5
	1970-79	-32	3	-42	-126	-71	-155	18	-53	-137	-95.0
	1980-82	-15	2	-27	-78	-40	-91	20	-20	-71	-45.5
	1960-82	-21	2	-26	-89	-45	-108	13	-95	-95	-63.5
Ghana	1962-69	-28	1	-65	-154	-92	-181	3	-89	-178	-133.5
	1970-76	-25	4	-60	-218	-81	-239	3	-78	-236	-157.0
	1977-84	n.a.	n.a.	n.a.	n.a.	n.a.	n.a.	n.a.	n.a.	n.a.	n.a.
	1962-76	-26	2	-63	-184	-87	-208	3	-84	-205	-144.5
Zambia	1960-70	n.a.	n.a.	n.a.	n.a.	n.a.	n.a.	n.a.	n.a.	n.a.	n.a.
	1971-79	-19	9	-80	-144	-90	-154	5	-85	-149	-117.0
	1980-84	-71	36	-304	-411	-339	-446	4	-335	-442	-388.5
	1971-84	-37	19	-160	-239	-178	-257	5	-173	-252	-212.5

(continued)

Table 2.2—Continued

		Total Price Transfers							Sum of Total Price and Nonprice Transfers		
				Output of Other Agricultural Products		Sum of Total Price Transfers					
Country	Period	Output of Selected Products (1)[a]	All Inputs (2)[b]	Assumption 1 (3)[c]	Assumption 2 (4)[c]	Assumption 1 (Sum of Columns 1, 2, & 3) (5)	Assumption 2 (Sum of Columns 1, 2, & 4) (6)	Nonprice Transfers (7)[d]	Assumption 1 (Sum of Columns 5 & 7) (8)	Assumption 2 (Sum of Columns 6 & 7) (9)	Average of Assumptions 1 & 2 (Half of Columns 8 & 9) (10)
		(percent of agricultural GDP)									
Group average[e]	1960s	-20.5	1.0	-37.5	-104.5	-57.0	-124.0	4.5	-52.5	-119.5	-86.0
	1970s	-25.3	5.3	-60.9	-162.7	-80.7	-182.7	8.7	-72.0	-174.0	-123.0
	1980s	-43.0	19.0	-165.5	-244.5	-189.5	-268.5	12.0	-177.5	-256.5	-217.0
	1960-84	-28.0	7.7	-83.0	-170.7	-103.3	-191.0	7.0	-96.3	-184.0	-140.2

Source: M. Schiff and A. Valdés, *A Synthesis of the Economics in Developing Countries*, vol. 4 of *The Political Economy of Agricultural Pricing Policy* (Baltimore, Md., U.S.A.: Johns Hopkins University Press, 1992).

Note: n.a. means not available.

[a] Column (1) refers to the change in the gross value of output of selected agricultural products as a result of direct price interventions (relative to the counterfactual simulation without intervention).

[b] Column (2) refers to transfers resulting from the price interventions on inputs, including credit subsidies (and replacing subsidies for rubber), on all agricultural products.

[c] Columns (3) and (4) refer to the additional effect of price-related transfers on the gross value of output for the rest of agriculture. Under assumption 1, the rate of nominal protection for the rest of agriculture is assumed to be zero (not shown); under assumption 2, it is assumed to equal nominal protection (or taxation) for the selected products.

[d] Column (7) transfers include public investment in irrigation, agricultural research and extension, and land improvements.

[e] Simple, unweighted group average of the three countries.

Broadly speaking, the issues related to economic policy reforms as they affect agriculture in LDCs can be usefully classified into four categories: (1) policy reforms to improve the economic environment for agriculture; (2) strengthening of the public sector to support technology development and transfer, education in rural areas, and infrastructure projects supportive of agriculture; (3) encouragement of opportunities for increased participation in the economy of the historically disadvantaged (that is, small farmers and landless workers); and (4) natural resources management.

In this paper the first of these four categories is addressed. This section is structured around three issues, namely, trade and macroeconomic factors, guidelines for successful agricultural price and trade reform, and the need for simultaneous reforms in selected sectors that impinge on the success of the agricultural trade policy reform process.

Trade and Macroeconomic Factors

These are perhaps the most important factors in the success or failure of agricultural price reform. The reduction in industrial protection alone would produce a major improvement in agricultural incentives, as indicated earlier in this chapter. For the sample of 18 countries analyzed (Schiff and Valdés 1992), just the reduction in industrial protection to, say, a uniform tariff of 15 percent would induce an increase in relative prices for agricultural tradables by approximately 22 percent relative to industry and by about 15 percent relative to the entire nonagricultural sector.

Moreover, there has been a strong interaction in the past between the macroeconomic circumstances and the prevalence of government controls of individual agricultural markets. For example, the majority of price controls have been imposed in an effort to reduce inflation, and price controls and quantitative restrictions have been intensified because of inflationary pressures or balance of trade difficulties or both. Thus, the persistence of macroeconomic disequilibrium will create strong pressures against the removal of price controls on farm products, particularly on food.

Much can be learned about the liberalization episodes and the role of the exchange rate in trade liberalization from the comparative study of 18 countries undertaken under the direction of Michaely, Choksi, and Papageorgiou (Shepherd and Langoni 1991). In presenting the findings of this study, Michaely concludes that

> A strong qualitative relationship is established between the behavior of the real exchange rate when liberalization is

> launched and the fate of the policy: the liberalization is likely to be sustained when the exchange rate increases and to collapse when it falls. An increase of the real rate appears to be almost a necessary condition for at least partial survival of a liberalization policy.

Michaely's analysis indicates that nominal devaluations accompanied by tight fiscal and monetary policies are almost invariably the policy actions required for real depreciations. A high exchange rate (higher than would be needed in the long term under a given tariff regime) will stimulate exports and will help contain imports at the beginning of the liberalization process. If there is one clear lesson from the experience of the bold trade liberalization programs in the Southern Cone countries in South America during the late 1970s (Corbo, Goldstein, and Khan 1987) and in New Zealand after 1984 (Sandrey and Reynolds 1990), it is the considerable risk for agriculture that can arise from the macroeconomic management of the economy. At the time, the financial strategy of the governments in these countries resulted in a very high real interest rate that attracted a considerable inflow of funds from abroad. As a result, the high interest rates, the high cost of capital, and the impact of higher capital inflows on the real exchange rate adversely affected agricultural investment and its international competitiveness and delayed the agricultural output response to the trade reform.

Today, unfortunately, practically the same appreciation phenomenon can be observed in Colombia and Peru; both countries initiated bold trade liberalization efforts in 1990/91, and the macroeconomic conditions have resulted in "cheap dollars" adversely affecting the competitiveness of the tradable sector just at the time when they are reducing their trade barriers. This will probably generate a strong additional resistance to the liberalization package from the pressure groups representing the import-competing sectors and will not get the support of those who otherwise would gain much, namely, the producers of exportables.

Thus, reduction of the indirect effect of fiscal deficit and real exchange rate appreciation and avoidance of sharp fluctuations in real interest rates and exchange rates are necessary and fundamental elements of a policy-reform package as it affects agriculture.

Guidelines for Successful Agricultural Price and Trade Reform

Four significant results have come out of recent studies on the agricultural trade regime in LDCs. First, there is a marked contrast between the direct policies adopted toward traditional export crops and those directed toward import-competing food products; governments heavily tax the production of exportables and protect the production of food. Second, quantitative restrictions (QRs) on agricultural trade (such as quotas, licenses, and state trading) are widespread in most LDCs. Third, a characteristic of the trade regime in farm products has been its discretionary and selective nature, its lack of transparency, and the implicit discrimination against subsectors of agriculture. Fourth, in some countries, revenues from trade taxes represent a significant share of government revenues; thus, a removal of trade distortions without increasing revenues from other sources might not be possible in those countries (see Tables 2.3 and 2.4).

Table 2.3—Principal trade and exchange rate reforms

Policy	Reform
Real Exchange Rate	Increase in real rate is a necessary condition for survival of trade liberalization policy.[a]
	The "fundamentals" are trade policy, overall level of expenditures (public and private), and foreign terms of trade.
	Nominal devaluation with tight fiscal and monetary policies is invariably required for a real devaluation. Nominal devaluation alone won't do it.
Trade	Reduction in industrial protection (tariffs and QRs)[b] will have a strong effect in raising the real exchange rate.
	The "exchange-rate protection" effect is strong. If the current average tariff equivalent on importables is, for example, reduced from 150 percent to 50 percent, the "true" protection to importables will be considerably higher than 50 percent after liberalization. The reduction of trade barriers increases the equilibrium exchange rate, providing protection to importables.
	The QR revolution. Removal of QRs on inputs and final goods in all sectors is the most fundamental trade and price reform.

[a] While the real exchange rate is not itself a policy instrument, the authorities can influence its level via policy changes on its underlying fundamentals (that is, government expenditures, level of protection, wage policy, and foreign capital flows).
[b] QR = quantitative restriction.

Table 2.4—Some guidelines on sequencing of trade and internal reforms

Trade Policy Reforms	Internal Reforms
Convert quantitative restrictions (QRs) to tariff equivalents in agricultural and nonfarm sector importables.	Promotion of competitiveness in factor and output markets including financial, agricultural marketing, transport, and communications. "Freedom of entry" is needed in all these activities.
Announce program of reduction of tariffs. Reach target level in not more than three years.	Fiscal reform is needed. Ideally, it should include value-added tax on all goods, domestic and traded; income tax reform; and reduction or elimination of trade taxes.
Reduce dispersion in tariffs, uniformly if possible, and not higher than 20 percent as target level.	A land market should be established. Success of policy reform requires capacity to adjust size, buy, sell, and rent land.
Remove subsidies on inputs and marketing and eliminate licensing beyond registration requirements.	
Reform the special case of agricultural price stabilization (associated with tariffication) to one or two basic styles.	

More transparency and fewer discretionary policies. It is submitted here that dismantling of QRs on input and final products, even if some degree of protection is maintained, is a condition of the liberalization package. Replacing QRs with tariffs has several advantages. First, the role of the price mechanism is enhanced. QRs are more selective and less visible than tariffs; they mask the level of protection and insulate the domestic markets from world market changes. Second, dismantling QRs could greatly reduce the role of state agencies in trade activities. Third, replacing QRs with tariffs generates government revenues, removing one of the obstacles to trade reform in some countries. Bold steps to eliminate QRs on output and inputs and dismantle the machinery to administer the QRs were an important element in the successful trade liberalization program in Chile and are an explicit component of the ongoing trade-reform programs in Argentina, Mexico, Brazil, Bolivia, Colombia, Venezuela, Peru, and several other countries.

A more neutral regime between exportables and importables. An important goal of trade reform is to achieve more neutrality in the trade regime, that is, narrowing the range of nominal and effective rates of protection. As documented earlier in the chapter, agricultural price interventions were found to have a strong antitrade bias, with a wide dispersion in tariff equivalences within importables as well as in the export tax equivalent on exportables. There are strong arguments against selectivity in the pattern of trade restrictions. One is a strictly economic argument. Many farm products are intermediate inputs in the processing industry, and depending on their share in the cost structure, even small differences in nominal tariffs across the economy can result in wide variations in effective rates of protection to processors, an unintended result of the reform.

Another argument derives from political-economy considerations. The experience with trade interventions suggests that through time a selective approach to interventions tends to be captured mostly by the more powerful pressure groups, deviating considerably from the initial motives for interventions. This has been the case, for example, with credit and input subsidies, captured mostly by the larger farmers, and with the protection to products produced in certain regions or by a certain class of farmers. Thus, liberalizing agricultural trade means not only lowering the average levels of protection and removing export taxation and restrictions, but also narrowing the range of nominal and effective rates of protection.

The special cases. Even though economic analysis identifies several economic motives for trade and price intervention in agricultural markets, the cases for deviations from the uniform tariff rule are very few (Valdés and Siamwalla 1988). These include the optimum tariff case and the fiscal revenue motive (both mentioned earlier), interventions to deal with world price instability, and food subsidies for the most vulnerable households. To avoid a capricious and distorting pattern of trade intervention, the goal should be in the direction of equality of nominal rates of protection on inputs and final products throughout the economy, including agriculture. Special cases should be exceptions, where the burden of proof is to demonstrate the merits of the special case.

Fiscal revenue and direct price interventions. When revenues from agricultural trade taxes are significant, ideally the reform process should include a fiscal reform to replace some of the trade taxes for other more neutral sources of revenues. This is clearly a relevant option in the near future for middle-income countries, but it might take longer to develop the institutional setup in low-income LDCs. One should consider, however, that one effect of tariffication (by replacing QRs) will be an increase in fiscal revenues, and thus the merits of explicit export taxes as source of revenues decline.

Price interventions and social objectives. Price and trade policies cannot protect all consumers. Faced with obvious fiscal constraints, most governments try to reduce the fiscal burden of subsidies by transferring part of the cost to producers of farm products through price controls, export quotas, and prohibitions. However, in most countries direct price interventions have had a small effect on household income. The findings in Schiff and Valdés 1992 suggest that the principal sources of the subsidy to consumers of food and raw materials are the indirect price interventions, basically the exchange-rate misalignment. Price interventions had a relatively larger real-income effect on farm producers than on urban consumers, transferring income from rural to urban households. The paradox is that in most countries the bulk of the poor are in rural areas; thus, in the long run, the net effect of price interventions may in many cases have been regressive. If taxing agriculture reduces rural demand for labor, rural employment and real wages will fall, leading to increased migration to the cities and increased competition for employment, and so to a fall in income in the informal urban sector as well. Agricultural price and trade policies are inefficient instruments to deal with social objectives. Targeted transfers to the poorest households should be chosen to replace overall price interventions.

Need for Simultaneous Reforms in Related Sectors

A variety of markets are subject to controls of varying degrees of severity. These include financial and labor markets, transport, and communications, the importance of which will vary from country to country. As is the case today with Eastern European agriculture, delineating the exact boundaries for a successful microeconomic reform package is obviously a very complex issue on which experience is limited.

It is widely recognized that the cost of adjustment (in terms of unemployment and financial pressure for farmers) precedes the benefits of liberalization and trade reform. There is, however, a real risk that the benefits in terms of agricultural output, employment, and farm income could take many years, reducing the political support for the reform. Due to the biological nature of agriculture, some adjustment lags are inevitable. However, the challenge is to identify the possible bottlenecks that could slow down the output response.

The following measures are believed to be particularly significant to consider in the sequence of reforms: (1) security of property rights and deregulation in land markets with respect to rentals, both of which are very important, at least in Latin America; (2) development of medium- and long-term credit lines with competitive interest rates and methods for dealing with accumulated farm debt; (3) development of competitive services on transport and communications, which is particularly important for the growth of nontraditional exports; and (4) public sector reforms to improve productivity in the use of public sector resources, and privatization of state agencies whose continued holding by the public sector is not justified on policy grounds.

Experience in the analysis of agricultural policy reforms in LDCs is still very limited. Further studies should be able to offer more precise guidelines about the order in which reforms must be undertaken. For example, should agricultural trade and price policy reforms follow stabilization and not be attempted before macroeconomic equilibrium and stability are firmly established? Similarly, should changes in agricultural trade and price policy reform, which could occur rapidly, be delayed because of others that would take longer (such as improving physical infrastructure, providing security of property rights, and developing an efficient service sector)? Several years will be needed to put these elements in place. It is important to move ahead to initiate the necessary reforms. The only case for delaying agricultural trade and price policy reform would seem to be for countries suffering unsustainable macroeconomic policies accompanied by high and variable rates of inflation, and by variability in the key macroeconomic variables, namely, the real exchange rate and real interest rate. A bold trade reform implemented before greater macroeconomic control is achieved would likely lead to the destabilization and eventual abandonment of the reform efforts.

APPENDIX: ON THE CONCEPT AND MEASURE OF AGRICULTURAL INCENTIVES

In their analysis of agricultural incentives, a feature of the approach in the study by Krueger, Schiff, and Valdés (1988) was the distinction between direct (or sectoral) and indirect (or economywide) price interventions. Agricultural incentives were defined as the domestic price of agricultural goods relative to the price of nonagricultural goods (P_a/P_{na}). Price interventions were then measured as the percentage departure from the relative price of agricultural goods that would have prevailed in the absence of sectoral price interventions as well as in the absence of trade interventions in the nonfarm sector and corrected for the exchange-rate misalignment (departure of the official exchange rate from the equilibrium exchange rate). For a given farm product, a negative price intervention occurred whenever the price of that good relative to the nonagricultural sector appeared below its counterpart price under a nonintervention scenario. Broadly, the nonintervention price is efficient under the assumptions that (1) the country is a pricetaker in the product in question, and (2) there are no externalities or economies of scale in production.

The direct (sectoral) price interventions were measured by the direct nominal protection rate, which is defined as the proportional difference between the relative domestic price and the relative border price of agricultural tradables (the effective rates of protection were also computed when data was available). The border prices, evaluated at the official exchange rate, are adjusted for quality differences and transport and storage costs; they are used as reference prices (that is, the domestic price that would have prevailed under free trade).

Policies underlying direct price interventions include tariffs and quotas, prior import licenses, direct price controls, taxes, and subsidies on products and inputs often operating through the activities of parastatals involved in the marketing of these products.

The indirect price interventions were defined as those arising from macroeconomic policies and trade policies on nonagricultural commodities, resulting in an exchange-rate misalignment.[4] Let P_a denote the price of agricultural tradables, P_{nat} the price of tradables outside agriculture, and P_{nah} the price of home goods outside agriculture. There are three economywide effects: higher industrial protection causes

---

[4] Exchange rate misalignment is defined as the departure of the official exchange rate from the equilibrium exchange rate.

P_a/P_{nah} to fall, that is, the real exchange rate appreciates; expansionary macroeconomic policies lead to a further appreciation of the real exchange rate and to a further fall in P_a/P_{nah}; and protection policies of industrial goods cause the reduction of P_a/P_{nat}. Defining $P_a/P_{na} = P_a/[\alpha P_{nat} + (1 - \alpha)P_{nah}]$ as agriculture's terms of trade (where α is the share of tradables in the nonagricultural sector), then the indirect effect of economywide policies was measured as the weighted average of the effects on P_a/P_{nah} and on P_a/P_{nat}.

The sum of the direct (*NPRd*) and indirect (*NPRi*) nominal protection rates was defined as the total nominal protection rate (*NPRt*), which measures the joint impact of sectoral and economywide policies.

REFERENCES

Corbo, V., M. Goldstein, and M. Khan. 1987. *Growth-oriented adjustment programs*. Washington, D.C.: International Monetary Fund/World Bank.

Honma, M., and Y. Hayami. 1986. The determinants of agricultural protection levels: An econometric analysis. In *The political economy of agricultural protection*, ed. K. Anderson and Y. Hayami, 39-49. London: Allen and Unwin.

Krueger, A. O., M. Schiff, and A. Valdés. 1988. Agricultural incentives in developing countries: Measuring the effect of sectoral and economywide policies. *World Bank Economic Review* 2 (3): 255-271.

Panagariya, A., and M. Schiff. 1990. *Commodity exports and real income in Africa*. World Bank PRE Working Paper WPS 537. Washington, D.C.: World Bank.

Sandrey, R., and R. Reynolds, eds. 1990. *Farming without subsidies: New Zealand's recent experience*. Wellington: Ministry of Agriculture and Fisheries.

Schiff, M., and A. Valdés. 1992. *A synthesis of the economics in developing countries*. Vol.4, *The political economy of agricultural pricing policy*. Baltimore, Md., U.S.A.: Johns Hopkins University Press.

Shepherd G., and C. G. Langoni, eds. 1991. *Trade reforms—Lessons from eight countries*. San Francisco, Calif., U.S.A.: International Center for Economic Growth.

Valdés, A., and A. Siamwalla. 1988. Foreign trade regime, exchange rate policy, and the structure of incentives. In *Agricultural price policy for developing countries*, ed. J. W. Mellor and R. Ahmed, 103-123. Baltimore, Md., U.S.A.: Johns Hopkins University Press for the International Food Policy Research Institute.

3
Agricultural Trade between Malawi, Tanzania, Zambia, and Zimbabwe: Competitiveness and Potential

Ulrich Koester

Regional integration among African countries has received increasing attention in recent years. Policymakers have accepted that trade could contribute in a major way to development and food security. The objective of this chapter is to address this issue by investigating the potential for intraregional trade in southeastern Africa, namely, between Malawi, Tanzania, Zambia, and Zimbabwe.

Economic theory suggests that a potential for trade exists if price ratios in the pretrade situation differ, if the set of products on the market differs, or if economies of scale in production exist. In short, the scope for trade expansion often depends on the dissimilarities of the countries in the pretrade situation. The dissimilarity emanates mainly from factor endowment, including natural resources, as well as from patterns of production, trade, and consumption. Prices, comparative advantage for individual commodities, overvaluation of currencies, policy environment, and instability in production and consumption of the main staples also contribute to the differences in the economic environment between the four countries. These dissimilarities are investigated in the first section of this report in order to assess whether a potential for trade expansion exists. Obstacles to trade expansion are explored in the next section, and, finally, policies to exploit the trade potential are discussed.

THE POTENTIAL FOR TRADE

Similarities and Dissimilarities of the Four Countries

The Factor Endowment and Ecological Conditions. Differences in comparative advantage, a main determinant of trade, are partly related to each country's endowment with factors of production, including the natural-resource base. Measuring these differences is limited by the availability of accurate assessment of natural conditions in these countries; nevertheless, some indicators have been computed and offer a basis for comparison among the four countries.

National resource base. The data in Table 3.1 reveal that there are significant differences in land-resource endowment. Malawi is by far the least-endowed country, yet 39 percent of cultivable land was unused in 1980. However, in 2010 this figure is projected to decline to 23 percent.

For Zambia, reserves of potentially cultivable land equaled nearly nine times the cultivated land in 1980. It is projected that even in 2010 there will still be a huge amount of unused cultivable land. The land resources of Tanzania and Zimbabwe fall between those of Malawi and Zambia.

Two messages can be drawn from these data. First, there remains an unexploited production potential in all four countries.[1] This even holds true with the present technology applied. Second, the land constraint differs significantly among countries. Zambia has a comparative advantage in moving to more land-tied agricultural production.

The land distribution is also different among the four countries. Even though Malawi has the smallest land area, it allocates a much larger share of it to crops than do the other three countries. Tanzania and Zambia reserve nearly 50 percent of their land area to pasture,

Table 3.1—Cropland per capita in Malawi, Tanzania, Zambia, and Zimbabwe in 1980 and projected for 2010

	1980		2010	
	Annual and Permanent Cropland	Reserves of Potential Cultivable Land	Projected Cropland Demand	Remaining Reserves of Cultivable Land
	(hectares)			
Malawi	0.42	0.27	0.20	0.06
Tanzania	0.48	1.44	0.29	0.36
Zambia	0.95	8.18	0.41	2.83
Zimbabwe	0.56	1.62	0.25	0.49

Source: Food and Agriculture Organization of the United Nations, *African Agriculture, The Next 25 Years: Atlas of African Agriculture* (Rome: FAO, 1986).

[1] This potential remains to the extent that inputs other than land are available.

while Malawi and Zimbabwe allocate less than 20 percent to that purpose (FAO 1989b).

Agroclimatic conditions. Malawi shows high agroclimatic suitability for maize, cassava, phaseolus beans, and sorghum, but not for millet. Zambia shows a definite advantage in suitability for all these crops except cassava. The percentages are lower for Tanzania and Zimbabwe for all five crops but especially for cassava. The diversity in crop suitability among the four countries indicates that the countries differ in their natural advantage to grow the five crops. This may translate into differences in production patterns. Tanzania, Zambia, and Zimbabwe would be better suited for maize, phaseolus beans, sorghum, and millet than for cassava. Zimbabwe is the least suited for producing cassava (Table 3.2).[2]

The variation in suitability for rainfed food crops is pronounced among adjoining regions of the four neighboring countries (Koester 1990). The suitability assessment for maize and phaseolus beans is very similar between Zambia, Malawi, and Tanzania across borders. On the other hand, there is more variation in the level of suitability of millet; Zambia's very suitable regions adjoin Zimbabwe's and Malawi's less suitable regions. Zambia's very suitable regions for cassava adjoin Malawi's less suitable regions (Koester 1990). This diversity across borders based on agroclimatic suitability could translate into variations in the production patterns of adjoining regions and therefore increase the potential for intraregional trade.

Reliability of production. It is well known that agricultural production is subject to drought in the region. However, the reliability of production outcome differs significantly on the subregional level. The data reveal that the four countries and the subregions in these countries differ in their production risk. The probability of production shortfall is only marginal in Malawi, but significant in Zimbabwe (Table 3.3). This supports the hypothesis that fluctuations in production may differ on the national and subregional level among the countries, and that there is potential for trade flows.

This potential will be even greater if changes in the climatic conditions vary across countries and subregions. Looking at the countries individually, some variability can be seen in annual rainfall

---

[2] Note that in computing these areas, no account is taken of fallow period requirements or of nonarable land requirements (that is, rangeland and land for forestry, urban areas, transportation, recreation and wildlife, reservoirs, and surface mining).

Table 3.2—Agroclimatic suitability assessment for rainfed production

Crop	Malawi	Tanzania	Zambia	Zimbabwe
		(percent of total land)		
Maize				
Very suitable	83	53	84	42
Suitable	17	23	13	17
Marginally suitable	0	5	2	9
Not suitable	0	19	1	32
Cassava				
Very suitable	24	9	0	0
Suitable	72	41	60	24
Marginally suitable	4	10	29	18
Not suitable	0	40	11	58
Phaseolus beans				
Very suitable	47	50	77	39
Suitable	52	25	19	20
Marginally suitable	0	8	3	13
Not suitable	1	17	1	28
Sorghum				
Very suitable	42	47	70	40
Suitable	58	26	26	19
Marginally suitable	0	7	4	18
Not suitable	10	20	0	23
Millet				
Very suitable	0	31	35	26
Suitable	53	43	59	27
Marginally suitable	17	8	1	27
Not suitable	30	18	6	20

Source: Author's calculations based on data from Food and Agriculture Organization of the United Nations, *African Agriculture, The Next 25 Years: Atlas of African Agriculture* (Rome: FAO, 1986).

Notes: The suitability classification measures the land area capable of yielding a percentage of the maximum attainable yield: very suitable, more than 80 percent; suitable, 40-80 percent; marginally suitable, 20-40 percent; and not suitable, less than 20 percent.

among the subregions. The coefficients of variation in rainfall in Malawi, Tanzania, Zambia, and Zimbabwe are 28, 44, 16, and 29 percent, respectively. The variability in annual rainfall among subregions is pronounced in Tanzania, moderate in Malawi and Zimbabwe, and slight in Zambia. In addition, Malawi, Tanzania, and Zimbabwe

Table 3.3—Agroclimatic suitability index

Country	Province	Index
Malawi	Northern	0.97
	Central	1.00
	Southern	0.91
Tanzania	Tasora	0.82
	Arusha	0.39
	Nbeya	0.87
	Dar es Salaam	0.89
	Mtwara	0.94
	Morogora	0.80
Zambia	Northern	0.56
	Luapula	0.50
	Eastern	1.00
	Central	0.92
	Southern	0.95
	Copperbelt	0.90
	North Western	0.90
	Western	0.93
Zimbabwe	Mashonaland West	0.91
	Mashonaland Central	0.91
	Mashonaland East	0.91
	Midlands	0.96
	Matabeleland North	0.41
	Matabeleland South	0.41
	Manicaland	0.92
	Masvingo	0.65

Source: Author's calculations and Southern African Development Coordination Conference, Regional Food Security Programme, "Regional Food Reserve," main report of a prefeasibility study prepared by Technosynesis, Zimbabwe, 1984.

Note: The index measures the ratio of actual production to expected production.

show very little correlation between actual and normal rainfall, while Zambia shows a high correlation. These coefficients seem to indicate that rainfall is more predictable and better distributed in Zambia than in the other three countries (Koester 1990).

An interesting aspect of the rainfall analysis is revealed in the very low correlation in monthly actual rainfall between adjacent subregions. The four countries under study belong to the subhumid and semiarid regions in Africa and have historically produced very similar crops. However, the coefficients of correlation computed between adjacent subregions show that except for the adjacent subregions of Karonga in Malawi and Iringa in Tanzania, which have a high coefficient of .52, most adjacent subregions have coefficients below .20 (Koester 1990). These coefficients suggest that rainfall patterns vary greatly among adjacent subregions and introduce the possibility that these subregions could specialize in different rainfed crops with variable requirements and, even more important, that the variations in production may offset each other on the subregional level.

Production Patterns in the Four Countries

Dissimilarities in the Production Patterns. In the absence of government interference, the present production patterns would be a reflection of comparative advantage. As it is, the present production patterns are the consequence of both comparative advantage and the revealed preferences of policymakers. Therefore, a comparison of the production patterns that indicates dissimilarities between countries may still serve as a preliminary indication of the scope for division of labor.

Production similarity indexes in Table 3.4 indicate a fairly high degree of dissimilarity, especially for the pairs Zimbabwe-Tanzania, Zambia-Tanzania, and Malawi-Tanzania. In contrast, the agricultural sectors of Zambia and Zimbabwe seem to be similar.

Concerning the production of the main staple foods—maize, cassava, phaseolus beans, sorghum, and millet—it is of interest to investigate whether the pattern of production reflects the countries' suitability for production. Table 3.5 contrasts the share of a country's suitability for production of the individual staples with the country's share in production. The data for Malawi and Tanzania, for example,

Table 3.4—Production similarity indexes, 1985-87

Country	Zambia	Zimbabwe	Tanzania
Malawi	60.41	67.57	31.98
Zambia	...	69.63	34.18
Zimbabwe	...	...	23.67

Source: Author's calculations based on the data for 60 agricultural products from Food and Agriculture Organization of the United Nations, "Production Yearbook Tape, 1988," FAO, Rome, 1989.

Note: An index of 100 shows that the production patterns of the two countries are completely similar, while a value of zero implies that the production patterns are completely dissimilar.

Table 3.5—Suitability shares and production shares for the main food staples, 1986-88

Crop	Malawi		Tanzania		Zambia		Zimbabwe	
	A	B	A	B	A	B	A	B
	(percent)							
Maize	100	83	76	27	97	81	59	81
Cassava	96	10	50	62	60	16	24	4
Phaseolus beans	99	5	75	3	96	0	59	2
Sorghum	100	1	75	5	86	2	59	5
Millet	53	1	74	3	94	2	53	7

Source: Author's calculations based on Food and Agriculture Organization of the United Nations, "Production Yearbook Tape, 1988," FAO, Rome, 1989; and Table 3.2.

Note: "A" is the suitability share (suitable and very suitable); "B" is the production share of these five staples.

illustrate the difference in these two countries' capacity to adjust their pattern of production. Malawi could produce, equally well, four out of the five main staples. Hence, depending on the set of incentives, production may shift away from maize in favor of other commodities such as cassava and beans. Tanzania is not generally as well suited as Malawi for producing the main staples, but the suitability does not vary much among the products. Thus, Tanzania also enjoys a high potential to adjust its pattern of production. Tanzania might be better off producing less cassava, while Malawi might be better off producing more cassava.

On the pattern of agricultural trade, the concentration of exports is high; during 1984-87, the three main agricultural export commodities made up 86 percent of all agricultural exports for Malawi, 76 percent for Zambia, and 73 percent for Tanzania and Zimbabwe. However, the composition of products within each group is partly similar for Malawi, Zambia, and Zimbabwe, while Tanzania has a very different export pattern (Table 3.6).

Calculations of export similarity indexes (Table 3.7), which correspond to the production similarity index, reflect the differences in production patterns. Especially the export structure of Tanzania seems to be quite different from that of the other three countries. But high values (for example, for the pairs Malawi-Zambia and Zambia-

Table 3.6—Export performances of main agricultural products, 1970-73, 1977-80, and 1984-87

Country/Product	Year	Export Value as Share of Total Agricultural Exports	Revealed Comparative Advantage Index	Comparative Export Performance Index
Malawi				
Tobacco, unmanufactured	1970-73	50.7	0.6	22.4
	1977-80	56.1	2.1	29.7
	1984-87	56.8	2.7	31.7
Tea	1970-73	22.3	3.6	20.9
	1977-80	19.4	5.2	19.1
	1984-87	19.7	5.3	18.4
Sugar and honey	1970-73	2.0	-0.7	0.4
	1977-80	13.2	1.8	2.3
	1984-87	9.0	2.2	1.9
Oilseeds, nuts	1970-73	11.5	4.0	2.5
	1977-80	6.3	3.7	1.4
	1984-87	2.2	a	0.5
Vegetables, fresh	1970-73	3.4	1.6	1.0
	1977-80	1.5	1.3	0.4
	1984-87	2.6	1.7	0.6
Zambia				
Tobacco, unmanufactured	1970-73	52.9	a	23.4
	1977-80	37.9	a	20.1
	1984-87	32.0	6.0	17.9
Maize	1970-73	11.8	-0.3	3.1
	1977-80	29.7	0.0	6.2
	1984-87	6.7	-1.5	1.8
Cotton	1970-73	8.0	3.8	1.7
	1977-80	18.4	4.1	5.1
	1984-87	12.4	4.2	4.2
Oilseeds, nuts	1970-73	14.4	6.2	3.2
	1977-80	7.3	1.4	1.6
	1984-87	9.2	2.2	2.2
Sugar and honey	1970-73	0.0	-5.3	0.0
	1977-80	2.7	1.6	0.5
	1984-87	32.0	4.4	6.7

(continued)

Table 3.6—Continued

Country/Product	Year	Export Value as Share of Total Agricultural Exports	Revealed Comparative Advantage Index	Comparative Export Performance Index
Zimbabwe				
Tobacco,	1970-73	39.5	[a]	17.4
unmanufactured	1977-80	34.6	1.9	18.3
	1984-87	47.5	2.3	26.5
Cotton	1970-73	12.7	[a]	2.6
	1977-80	17.6	3.4	4.9
	1984-87	16.9	6.5	5.8
Meat, fresh	1970-73	17.5	[a]	2.3
	1977-80	10.9	1.0	1.5
	1984-87	4.4	0.2	0.6
Sugar and honey	1970-73	9.2	1.7	1.7
	1977-80	8.8	2.8	1.5
	1984-87	8.7	2.8	1.8
Maize	1970-73	9.9	[a]	2.5
	1977-80	7.5	-1.1	1.6
	1984-87	5.6	-1.2	1.6
Tanzania				
Coffee	1970-73	22.7	4.0	4.3
	1977-80	40.8	10.0	6.0
	1984-87	51.3	[a]	8.9
Cotton	1970-73	18.8	5.0	3.9
	1977-80	13.6	5.8	3.8
	1984-87	13.0	[a]	4.4
Spices	1970-73	12.6	2.9	25.5
	1977-80	9.3	2.6	18.9
	1984-87	8.6	3.7	13.1
Fruit, fresh	1970-73	10.0	2.9	2.0
	1977-80	6.8	3.6	1.5
	1984-87	3.7	8	0.7
Vegetable	1970-73	10.8	4.5	30.9
fibers	1977-80	7.1	8.1	34.3
	1984-87	2.4	[a]	10.6
Region				
Tobacco,	1970-73	25.0	1.9	11.0
unmanufactured	1977-80	26.8	3.0	14.2
	1984-87	37.8	3.2	21.1

(continued)

Table 3.6—Continued

Country/Product	Year	Export Value as Share of Total Agricultural Exports	Revealed Comparative Advantage Index	Comparative Export Performance Index
Region (continued)				
Coffee	1970-73	10.8	2.6	2.0
	1977-80	18.0	3.8	2.7
	1984-87	17.3	5.4	3.0
Cotton	1970-73	14.3	4.6	3.0
	1977-80	12.9	3.4	3.6
	1984-87	12.1	3.9	4.1
Tea	1970-73	4.8	1.0	4.5
	1977-80	6.9	2.6	6.8
	1984-87	7.6	5.8	7.1
Sugar and honey	1970-73	3.9	-0.7	0.7
	1977-80	6.5	0.6	1.1
	1984-87	7.3	1.1	1.5

Source: International Food Policy Research Institute, "Inter-LDC Trade Data Base," IFPRI, Washington, D.C., 1987.

Notes: The revealed comparative advantage (RCA) index compares the share of product i in the total agricultural exports with the share of this product in total agricultural imports. The higher the RCA index, the more successful a country is in exporting the product in question. RCA approaches positive infinity as imports go to zero. The comparative export performance (CEP) index is analogous to the comparative production performance index. The respective formulas are

$$RCA = \ln(X_i/M_i : \sum_{i=1}^{49} X_i \ / \ \sum_{i=1}^{49} M_i) \text{ , and}$$

$$CEP = X_i/X_{iw} : (\sum_{i=1}^{49} X_i \ / \ \sum_{i=1}^{49} X_{iw}) \text{ ,}$$

where X_i and M_i denote exports and imports, respectively, of 49 agricultural products and X_{iw} is world exports of product i.

[a] Values are undefined because imports of the product in question equal zero.

Zimbabwe for the period 1984-87) do not support the view that there is no potential for regional trade. Finger and Kreinin (1979) found similarity indexes around 50 for the United States and the European Community (EC) export patterns in the early 1970s, a period that was

Table 3.7—Export similarity indexes, 1970-73, 1977-80, and 1984-87

Year/Country	Zambia	Zimbabwe	Tanzania
1970-73			
Malawi	70.59	52.33	19.76
Zambia	...	60.81	21.49
Zimbabwe	...	...	24.12
1977-80			
Malawi	49.01	50.68	17.43
Zambia	...	66.05	23.39
Zimbabwe	...	...	32.75
1984-87			
Malawi	52.40	69.28	20.87
Zambia	...	65.78	27.60
Zimbabwe	...	...	32.43

Source: Author's calculations based on 49 agricultural products. Export data were taken from United Nations Conference on Trade and Development, "Export Value Tape," Geneva, 1989.

Note: An index of 100 means that the exports of the two countries are completely alike; an index of 0 means that they are completely dissimilar.

followed by considerable expansion in EC-U.S. trade. Comparison of the different time periods does not show a clear tendency to more similarity or dissimilarity. The large values for Malawi, Zambia, and Zimbabwe are mainly due to export of unmanufactured tobacco, which is the main agricultural export commodity of each of the three countries (Table 3.6). The share of unmanufactured tobacco in total agricultural exports increased from 35 percent in 1977-80 to 48 percent in 1984-87 for Zimbabwe. This explains the increase in the similarity of the export pattern of Malawi and Zimbabwe; thus, the export similarity index increased from 51 in 1977-80 to 69 in 1984-87.

Instability in Production. Based on the differences in the ecological conditions in the individual countries, one would expect that instability in production of the main staples would differ among the countries under investigation.

Calculations for the four countries for the period 1961-88 (Table 3.8) reveal that production is fairly volatile for total grains and individual types of grains. However, there are significant differences between types of grains and between countries. Maize is the most important cereal in the region, accounting for 78 percent of total cereal

Table 3.8—Instability in cereal production in four countries of the Southern African Development Coordination Conference, 1961-88

Cereal/Country	Production Share	Coefficient of Variation	Correlation Coefficients Malawi	Zambia	Zimbabwe
Wheat					
Malawi	0.00	0.46	1.00	...	...
Zambia	0.04	0.75	-0.11	1.00	...
Zimbabwe	0.61	0.31	-0.35	0.23	1.00
Tanzania	0.35	0.25	-0.08	-0.26	0.13
Region	1.00	0.22	...	...	...
World	...	0.04	...	...	...
Rice					
Malawi	0.13	0.52	1.00	...	...
Zambia	0.01	0.54	0.25	1.00	...
Zimbabwe	0.01	0.68	-0.16	-0.10	1.00
Tanzania	0.86	0.28	0.20	-0.56	-0.16
Region	1.00	0.26	...	...	...
World	...	0.03	...	...	...
Maize					
Malawi	0.24	0.11	1.00	...	...
Zambia	0.20	0.24	0.11	1.00	...
Zimbabwe	0.30	0.35	0.23	0.38	1.00
Tanzania	0.26	0.22	-0.45	0.00	-0.05
Region	1.00	0.14	...	...	...
World	...	0.07	...	...	...
Millet/sorghum					
Malawi	0.08	0.42	1.00	...	...
Zambia	0.10	0.34	0.35	1.00	...
Zimbabwe	0.29	0.24	-0.14	0.23	1.00
Tanzania	0.52	0.30	0.34	-0.03	-0.27
Region	1.00	0.18	...	...	...
World	...	0.07	...	...	...
Total cereals					
Malawi	0.20	0.12	1.00	...	...
Zambia	0.17	0.23	0.29	1.00	...
Zimbabwe	0.30	0.31	0.26	0.43	1.00
Tanzania	0.33	0.19	-0.13	0.22	-0.06
Region	1.00	0.12	...	...	...
World	...	0.03	...	...	...

Source: Author's calculations based on data from Food and Agriculture Organization of the United Nations, "Production Yearbook Tape, 1988," FAO, Rome, 1989.

production. Malawi, a relatively stable producer of maize, is the only country where maize production is less volatile on the country level than on the regional level. Apart from this special case, market integration could contribute to more stable supply on the country level. Tanzania has a relatively stable cereal production; therefore, trade would contribute less to stabilizing internal supply. Stability could also be improved if countries were to adjust their production pattern by emphasizing those types of grain for which they are more stable than the partner countries.

A trade potential based on fluctuations in production becomes even more evident if not only the national level is considered, but also the subregional level. Based on the differences in the suitability indexes presented above, one can expect production instability to vary significantly across subregions. Koester (1986) even found that neighboring subregions located on different sides of national borders suffer quite differently from fluctuations in total cereal production. However, the corresponding calculations may be based on somewhat unreliable data. Hence, some additional analysis including different types of grain is needed.

The Consumption Pattern and the National Balance of Production and Consumption. The potential for growth in intraregional trade is higher if countries with surplus production in some staples are bordered by countries with deficit production in the same staple. The data in Table 3.9 indicate that the pattern of consumption varies across countries. Maize is the dominant staple for Malawi, Zambia, and Zimbabwe. In addition to maize, cassava is an important product in the diet of people in Tanzania. This variance in the consumption pattern reflects the countries' production patterns and could lead to a significant trade potential. Cassava is much less prone to drought than the other main staples. Hence, a higher share of cassava in total production may be considered as a built-in stabilizer of production and consumption. It can therefore be expected that Tanzania will be less affected by shortfalls in cereal production than the other countries. Moreover, trade in cassava could help stabilize the region's consumption.

Scope for Expanding Intraregional Trade

Transborder trade could be used to meet three distinct tasks: to bridge the gap between national demand and national supply for individual products in years of normal harvest; to even out annual fluctuations in production; and to even out subregional deficits and surpluses.

Table 3.9—Degree of self-sufficiency in grains and grain consumption patterns, 1986-88

Country	Maize		Wheat		Rice		Millet and Sorghum		Cassava		Total	
	A	B	A	B	A	B	A	B	A	B	A	B
	(percent)											
Malawi	101.4	84.7	13.1	0.8	95.8	1.8	19.6	9.4	103.5	3.3	93.0	100
Tanzania	109.0	37.1	57.5	2.3	80.5	10.5	89.5	18.1	102.4	32.0	99.2	100
Zambia	137.6	81.9	29.3	7.0	110.8	0.8	91.3	4.7	100.0	5.6	125.5	100
Zimbabwe	131.4	69.4	69.9	16.4	2.6	0.8	112.4	12.1	100.0	1.2	117.6	100
Aggregate	117.9	55.7	59.7	5.4	79.7	6.2	86.8	14.1	102.3	18.6	105.1	100

Source: Author's calculations based on Food and Agriculture Organization of the United Nations, "Food Balance Sheets Tape," FAO, Rome, 1990.

Notes: A is the ratio of the average production in calorie-equivalents from 1986 to 1988 to the total domestic utilization for 1987. B is the share of grain consumption in total grain consumption in 1987.

Bridging the Gap between National Demand and National Supply through Redirection of Present Trade Flows. Given the present production and consumption patterns, trade among the countries could easily be increased if a redirection of trade flows were possible. It might well be that some of the countries import from outside the region, while other countries export the same product to outside the region.

Despite trade restrictions, the regional market between Malawi and the other three countries seems to be quite developed for some agricultural products (Tables 3.10 through 3.12). Some of Malawi's agricultural exports for 1981-84 flow exclusively to Zimbabwe (crude

Table 3.10—Malawi's exports of agricultural products to Zimbabwe as a percentage of Malawi's exports to the world, 1962-65 to 1981-84

Commodity	1962-65	1966-70	1971-75	1976-80	1981-84
			(percent)		
Crude rubber	100	100	0	0	100
Meat preparations	0	0	0	0	100
Fruit, preserved	0	100	62	48	100
Cocoa	0	0	39	0	100
Cereal preparations	0	17	42	79	97
Vegetables, fresh	16	12	13	1	70
Rice	73	78	60	26	70
Animal feed	97	48	34	15	62
Vegetables, roots	0	33	5	0	62
Food preparations	0	0	69	57	60
Spices	0	0	0	0	46
Fruit, fresh	0	100	0	66	44
Crude vegetables	0	13	22	4	31
Alcoholic beverages	0	0	0	0	31
Hides and skins	0	27	25	5	10
Oilseeds, oil nuts	0	0	0	1	9
Meal and flour of cereals	0	100	38	0	9
Animal and vegetable oils	0	6	0	100	7
Cotton	10	13	21	1	3
Crude animal matter	0	0	17	2	0
Other cereals	64	44	50	77	0
Meal and flour of cereals	0	67	50	0	0
Margarine and shortenings	0	0	88	0	0
Animal oils and fats	0	0	86	0	0
Wheat and meslin	0	72	0	0	0
Sugar confectionery	0	25	0	0	0

Source: International Food Policy Research Institute, "Inter-LDC Trade Data Base," IFPRI, Washington D.C., 1987.

Table 3.11—Malawi's exports of agricultural products to Zambia as a percentage of Malawi's exports to the world, 1962-65 to 1981-84

Commodity	1962-65	1966-70	1971-75	1976-80	1981-84
			(percent)		
Meal and flour of cereals	0	0	43	11	91
Tobacco, manufactured	0	100	75	0	83
Live animals	0	0	33	3	70
Maize	0	18	10	0	52
Crude vegetables	0	2	17	23	29
Rice	21	12	25	25	18
Animal feed	0	23	37	24	17
Food preparations	0	100	31	10	6
Fruit, fresh	0	0	100	0	1
Vegetables, fresh	4	7	5	4	1
Vegetables, roots	0	67	18	2	1
Alcoholic beverages	0	0	100	46	0
Sugar and honey	0	95	1	1	0
Cereal preparations	0	83	16	18	0
Fixed vegetables	100	100	100	0	0
Meat, fresh	100	100	0	0	0
Fruit, preserved	0	0	34	2	0
Wool and animal hair	0	0	0	100	0
Other cereals	0	0	50	23	0
Meal and flour of cereals	0	0	50	12	0
Margarine and shortenings	0	0	13	0	0
Sugar confectionery	0	50	0	0	0
Wheat and meslin	0	28	0	0	0

Source: International Food Policy Research Institute, "Inter-LDC Data Base," IFPRI, Washington, DC., 1987.

rubber, meat preparations, preserved fruit, and cocoa), Zambia (meal and flour, and manufactured tobacco), or Tanzania (cheese and curd). The same observation holds partially for Tanzania, which exports 100 percent of its fresh meat to Zambia (Table 3.13). Nevertheless, the pattern of the previous periods indicates that in the case of Tanzania, exports to Zambia lost in importance, particularly for maize, eggs, wheat and meslin, and dried fruit.

The existing trade pattern between these four countries and the rest of the world, as indicated in Table 3.14, shows the potential for developing intraregional trade through a redirection of the trade flows. For example, it is possible that out of the US$5.1 million in maize imports from Malawi to the rest of the world, US$2.4 million could

Table 3.12—Malawi's exports of agricultural products to Tanzania as a percentage of Malawi's exports to the world, 1962-65 to 1981-84

Commodity	1962-65	1966-70	1971-75	1976-80	1981-84
			(percent)		
Cheese and curd	0	0	0	100	100
Butter	0	0	0	100	71
Maize	0	0	22	0	18
Tobacco, manufactured	0	0	0	0	17
Food preparations	0	0	0	27	3
Live animals	0	0	0	43	0
Eggs	0	0	0	100	0

Source: International Food Policy Research Institute, "Inter-LDC Data Base," IFPRI, Washington, DC., 1987.

Table 3.13—Tanzania's exports of agricultural products to Zambia as a percentage of Tanzania's exports to the world, 1962-65 to 1981-84

Commodity	1962-65	1966-70	1971-75	1976-80	1981-84
			(percent)		
Meat, fresh	62	91	98	0	100
Vegetable fibres	0	0	1	5	0
Vegetables, fresh	0	6	1	1	0
Fixed vegetables	0	16	100	0	0
Other fixed vegetables	0	0	11	17	0
Alcoholic beverages	0	33	33	0	0
Margarine and shortenings	0	0	100	0	0
Tobacco, unmanufactured	7	0	0	0	0
Sugar and honey	83	2	6	0	0
Food preparations	0	100	3	0	0
Fruit, preserved	6	0	15	48	0
Rice	2	100	24	1	0
Meat preparations	0	5	6	13	0
Cereal preparations	0	0	100	0	0
Live animals	0	0	0	10	0
Crude rubber	0	0	0	86	0
Animal oils and fats	0	0	0	82	0
Maize	0	100	100	70	0
Vegetables, roots	0	50	10	0	0
Eggs	0	100	100	0	0
Wheat and meslin	0	100	100	0	0
Fruit, dried	0	100	100	0	0

Source: International Food Policy Research Institute, "Inter-LDC Data Base," IFPRI, Washington, DC., 1987.

Table 3.14—The pattern of agricultural trade for the region, 1981-84

	Zambia Imports Originating in			Zimbabwe Imports Originating in			Tanzania Imports Originating in		Malawi Imports Originating in		Rest of World Imports Originating in	
Commodity	Malawi	Tanzania	Rest of World	Malawi	Tanzania	Rest of World	Malawi	Rest of World	Tanzania	Rest of World	Malawi	Tanzania
						(US$1,000)						
Tobacco, unmanufactured	208	...	0	1,932	...	7	...	1	...	12	117,107	12,832
Sugar and honey	0	...	94	39	...	47	...	2,360	...	21	43,367	2,781
Tea	116	...	105	182	...	0	...	2	...	1	42,781	19,521
Maize, unmilled	8,675	...	2,403	0	...	11,255	2,949	28,186	...	562	5,051	...
Oilseeds, oil nuts	28	...	134	758	...	16	...	7	...	1	7,200	1,862
Vegetables, roots and tubers	49	...	50	2,723	...	136	...	127	...	37	1,632	...
Coffee	14	12	588	16	...	2	...	17	...	1	2,237	143,549
Rice	415	...	3,420	1,581	...	1,738	...	33,986	...	3,407	265	148
Animal feed	166	...	671	621	...	235	...	689	...	2	212	3,453
Cotton, raw	1	...	52	26	...	65	...	23	...	11	808	54,464
Hides and skins	0	...	0	74	...	0	...	49	...	10	693	2,661
Other fixed vegetable oils	0	...	125	4	...	4	...	465	...	16	584	229
Vegetables, fresh and frozen	3	...	165	167	...	1,858	...	613	...	72	68	8,210

(continued)

Table 3.14—Continued

Commodity	Zambia Imports Originating in			Zimbabwe Imports Originating in			Tanzania Imports Originating in		Malawi Imports Originating in		Rest of World Imports Originating in	
	Malawi	Tanzania	Rest of World	Malawi	Tanzania	Rest of World	Malawi	Rest of World	Tanzania	Rest of World	Malawi	Tanzania
						(US$1,000)						
Fruit, fresh, and nuts	3	...	26	103	...	28	...	94	...	23	127	60,511
Crude vegetable materials	64	...	721	70	...	421	6	917	...	84	83	1,940
Fruit, preserved	0	...	19	216	...	38	...	90	...	11	0	30
Spices	4	...	30	62	...	75	...	45	...	32	68	37,128
Meal and flour of cereals	118	...	0	12	...	323	...	291	...	52	0	545
Food preparations	5	...	920	46	...	216	2	1,930	...	355	24	874
Live animals	44	...	829	0	...	504	...	1,099	...	219	19	824
Vegetable fibers	...	76	7	0	...	19	...	20	...	4	...	20,733
Meat, fresh	...	3	41	0	...	471	...	211	...	29	...	0
Sum of above	9,913	91	10,400	8,632	...	17,458	2,957	71,222	...	4,962	222,326	372,295
Potential of additional—												
imports from Malawi	...	...	4,286	...		6,095	...	8,912	...	...	...	...
imports from Tanzania	...	...	4,583	...		3,998	...	...	...	1,075	...	...

Source: International Food Policy Research Institute, "Inter-LDC Data Base," IFPRI, Washington, D.C., 1987.

have been redirected to Zambia. A similar observation applies to oilseeds and oil nuts, coffee, animal feed, fixed vegetable oils, and fresh fruit and nuts.

The redirection of agricultural imports to Zambia would have resulted in a 43 percent increase in imports from Malawi, and a 5,036 percent increase from Tanzania. Similarly, Zimbabwe's imports from Malawi could have increased by 71 percent. Although Malawi and Zimbabwe did not import from Tanzania, there would have been a potential for imports amounting to US$1 million and US$4 million, respectively. Tanzania's imports from Malawi could have been increased by as much as 301 percent.

These data clearly support the view that the four countries under study could increase trade with each other just by redirecting trade flows. Furthermore, the widely held perception that Zimbabwe would become the dominant exporter in a more liberal trading environment and that other countries might not be able to penetrate the Zimbabwean market cannot be supported.

Indeed, intraregional trade, compared with extraregional trade, could become quite important for specific products, even with the present production and consumption pattern, which does not reflect the individual country's comparative advantage undistorted. The trade potential index (TP) informs on the importance of individual products for intraregional trade. The coefficient has been calculated as

$$TP = [Min(X_{i1}M_i)/Max(X_{i1}M_i)] \cdot 100.$$

This index measures the share of the region's exports (imports) of a specific commodity that are matched by imports (exports) of the same commodity. The results of the calculations for products with an index greater than 10 are shown in Table 3.15. Products with the highest overlap are live animals, maize, cereal preparations, vegetable roots, and crude vegetable materials. However, it should be noted that the index for the individual products is not constant over time. This is a reflection of the changing market conditions, which are susceptible to the weather. It is somewhat surprising that the region as a whole shows a decline of the index over time. Obviously, the countries have not adjusted their consumption and production pattern in order to take advantage of intraregional trade.

Because this index includes the present intraregional trade flows, it might be argued that the index overestimates the actual trade potential. But since market integration would certainly affect a country's

Table 3.15—Potential intraregional trade as a percentage of foreign trade, 1970-73 to 1984-87

Commodity	1970-73	1977-80	1984-87
		(percent)	
Live animals	69.60	36.26	95.86
Meat, fresh	26.12	2.80	8.39
Meat, dried	69.31	7.92	12.42
Meat preparations	24.08	8.23	3.94
Butter	4.77	6.44	23.52
Cheese and curd	0.59	37.85	13.44
Eggs	12.12	4.94	3.58
Rice	47.65	0.99	0.00
Barley	26.32	0.00	0.00
Maize	59.82	79.17	83.15
Cereals, others	57.29	15.60	35.87
Meal and flour of cereals	87.43	22.75	34.84
Cereal preparations	0.57	44.39	83.35
Fruit, dried	0.85	18.58	0.00
Fruit, preserved and prepared	17.63	86.23	63.42
Vegetables, fresh	43.96	14.29	24.49
Vegetables and roots, prepared	1.51	27.34	69.66
Sugar and honey	51.19	11.64	6.58
Sugar confectionery and preparations	0.83	70.01	13.16
Cocoa	58.63	30.08	17.26
Chocolate and other preparations	0.00	10.90	16.34
Animal feed	29.61	32.41	17.81
Margarine and shortening	19.83	52.91	50.35
Food preparations	0.00	0.00	50.00
Alcoholic beverages	31.80	5.21	8.81
Wool and animal hair	63.87	37.56	30.72
Crude vegetable materials	39.55	41.51	82.57
Fixed vegetable oils, soft	44.86	59.23	6.96
Other fixed vegetable oils	29.57	20.97	8.97
Region Total	11.01	7.39	5.95

Source: International Food Policy Research Institute, "Inter-LDC Trade Data Base," IFPRI, Washington, D.C., 1987.

Note: The table lists all products with a value greater than 10 in one of the periods.

production and trade pattern and the dynamic effects will boost intraregional trade, it must be emphasized that these results more likely underestimate the actual trade potential.

Trade Expansion Based on Subregional Surpluses and Deficits. It is a widely held view among African policymakers and even among some economists (Lipton 1988) that trade between the countries could only develop if there were a national surplus. However, this argument fails to appreciate that countries are not homogeneous units and that a country might be better off if it exports its subregional surplus to nearby subregions, which may be located in neighboring countries and may be in deficit. The potential for such transborder trade has been investigated by Nuppenau (see Chapter 12). Based on subregional surpluses and deficits in years of normal harvest, he estimates that transborder trade in maize could contribute to a significant saving in transport costs of at least 12 percent. However, this is certainly an underestimation, because he assumed as a point of reference a situation where internal trade would have been conducted in a cost-minimizing way, which is surely not the case. The present pattern of internal trade implies that public transport is provided to transport maize in its raw form immediately after harvest to the public stores and to the mills. Later in the year, the processed maize (that is, meal) is transported back to the regions. If transborder trade were permitted, maize would be directly transported from the surplus subregions to the deficit regions, but not necessarily within a few months after harvest. The reduction in the drain on the transport sector would be much more than indicated by Nuppenau.

Nuppenau also investigated the impact of a drought situation on intraregional trade flows. The main message is that trade flows would change their direction because individual subregions are differently affected by drought. Furthermore, the region as a whole could save in transport costs if the countries were to trade with each other, even if each individual country had to import from outside the region.

Trade Expansion Based on Supply Fluctuations. Agricultural production in the Southern African Development Coordination Conference (SADCC) region fluctuates significantly over time. Trade can contribute to stabilizing supply when national fluctuations in production are greater than regional fluctuations. Thus, free trade between the four countries could be an efficient substitute for national stockpiling and might be used to even out fluctuations in national production.

The deviation from trend in maize production in the four countries during 1970-89 is shown in Figure 3.1. In 16 of these 20 years, deviations from the trend are in opposite directions for the four countries. This supports the hypothesis that intraregional trade can be used to stabilize national supply. If, in contrast, countries pursued a national policy to stabilize supply without relying on trade, some countries would have to build up stocks in years when neighboring

Figure 3.1—Deviation from trend in maize production, 1970-89

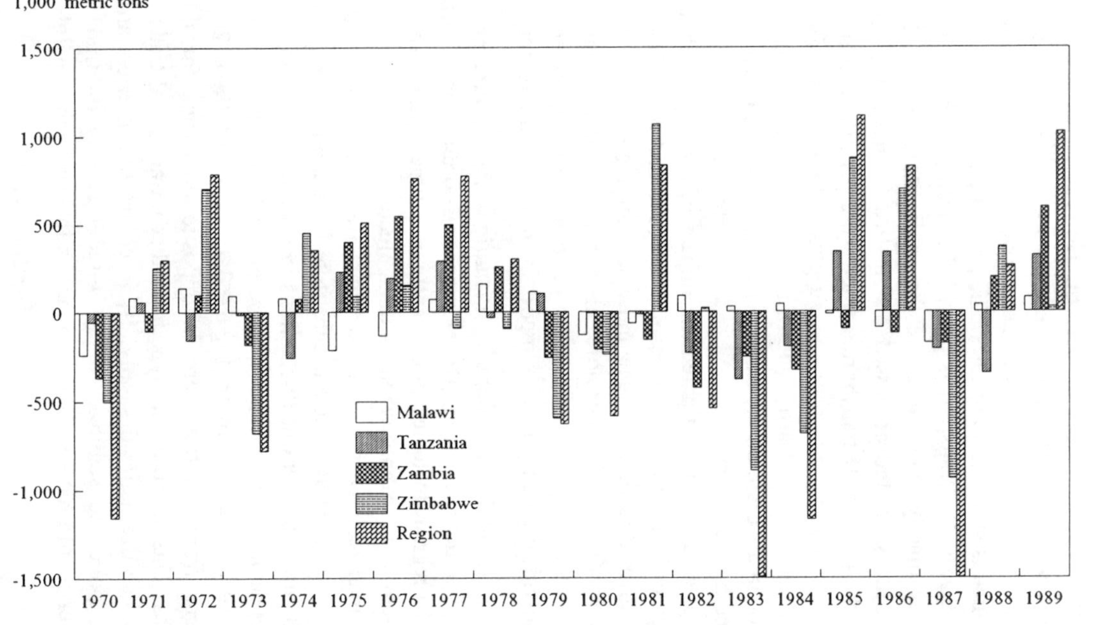

Source: Food and Agriculture Organization of the United Nations, "Production Yearbook Tape, 1988," FAO, Rome, 1989.

countries either depleted their stocks or had to increase their imports. Thus, intraregional trade based on fluctuations in production could help save storage and transport costs.

It is well known that the individual main staple foods are differently prone to drought. Cassava is the most drought resistant, followed by millet and sorghum. As the production pattern differs among the countries and as the composition of the individual products in the diet varies, too, trade would allow substitution in consumption and mitigate the effects of production shortfalls in specific commodities in specific countries. The expected benefits from substitution in consumption would be greater if the consumer price ratios differed among the countries. This aspect will be investigated in more detail below.

Induced changes in the consumption pattern could contribute even more in stabilizing food consumption because some of the staple is also fed to animals. Changes in price ratios between those that are preferred by humans and those that can be fed to animals as well would also induce trade and thus ease the strain on food security. The new development in the market for yellow maize could be exploited to serve this purpose.

Scope for Expanding Intraregional Trade by Adjusting Internal Production and Consumption Patterns

Market integration through liberalizing trade could affect the internal production and consumption patterns in several ways.

First, it is widely accepted that internal resources are not used efficiently. Not only are some sectors taxed and others subsidized, but the strong regulation of the internal economies ties up too many resources in the government and in parastatals. In addition, the private sector has to use more resources to overcome the red tape and to exploit the present system through lobbying and rent-seeking. Hence, liberalizing the economy would free resources for more productive activities by allowing price signals to determine the flow of factors in the economy.

OBSTACLES TO TRADE AND IMPLICATIONS FOR POLICIES REFORM

Extent of Present Taxation of the Agricultural Sector

The agricultural sector may be taxed due to price and macroeconomic policies. Price policies that set agricultural producer prices below export or import parity prices (evaluated at the official exchange rate) tax the agricultural sector directly. In addition, overvalued currencies indirectly tax tradables and consequently depress the agricultural sector further, since in most African countries a large share of traded commodities comes from agriculture. Moreover, the internal policies that directly subsidize other sectors increase the indirect taxation of the agricultural sector.

The agricultural sector may be distorted internally because of differences in price ratios between individual agricultural products due to discriminatory price controls. Adjusting these price ratios would lead to reallocation of resources within the sector and thus could open up a potential for intraregional trade.

Quiroz and Valdés (Chapter 4) estimated the total effects of government policies for the agricultural sector.[3] The results show that agriculture is taxed heavily in all four countries and that the protection (or taxation) differs among countries and across products (Chapter 4, Table 4.6). The conclusion is that a more liberal trading environment would lead to an outward shift in the agricultural sector's production-possibility curve. The countries that taxed their agriculture the most are Malawi and Zambia. The implications are very significant for Zambia, where a huge proportion of the potentially cultivable land is unexploited. However, the protection rates differ across products and, consequently, liberalization could also lead to a change in the production pattern. In Tanzania, liberalization would create incentives to expand export crops such as coffee and cotton, whereas rice and maize production would become less profitable. Malawi would have to institute an incentive system that would equally stimulate the export crops such as tobacco, wheat, and maize. Hence, the new patterns of production in Malawi and Tanzania could open up an additional trade potential for exports of maize from Malawi to Tanzania. Maize, which is highly taxed in Zambia, could be imported into Tanzania from Zambia.

The findings on Zimbabwe deserve special comment. Wheat and cotton are strongly linked in production, yet wheat is protected and cotton highly taxed. It may well be that even if wheat prices were lowered, wheat production would increase because of higher cotton prices.

Quiroz and Valdés (Chapter 4) could present only a somewhat aggregated view. Table 3.16 indicates how the maize economy might be affected by price policies. Maize producer prices differ widely

---

[3] See Chapter 2, Appendix, for details on the definitions of direct and total effects.

Table 3.16—Maize producer prices for Malawi, Tanzania, Zambia, and Zimbabwe, 1980-88

Year	Malawi	Tanzania	Zambia	Zimbabwe	World
			(US$/metric ton)		
1980	81	122	165	138	117
1981	74	181	173	173	122
1982	105	189	192	158	101
1983	94	197	162	119	127
1984	85	262	152	113	127
1985	71	300	116	111	105
1986	73	193	84	107	82
1987	67	128	73	108	71
1988	68	88	86	107	100

Source: Food and Agriculture Organization of the United Nations, "Producer Prices Tape," FAO, Rome, 1989; and International Monetary Fund, *International Finance Statistics* (Washington, D.C.: IMF, 1990).

Notes: Producer prices were converted to U.S. dollars at the official exchange rate. World prices are U.S. gulf ports f.o.b.

among Malawi, Tanzania, Zambia, and Zimbabwe. It should also be noted that prices have not moved in the same direction among the four countries or in tandem with world market prices. For example, world market prices increased from 1982 to 1983 and remained stable in the next year, but except for Tanzania, prices dropped significantly in these countries: by 19 percent for Malawi, 21 percent for Zambia, and 28 percent for Zimbabwe. In 1986 and 1987, the prices in Malawi and Zambia seem to approach the world market prices. From 1980 through 1985, Tanzania's maize prices increased to an extremely high level. In 1985, they were nearly three times as high as prices in Zambia and Zimbabwe and more than four times as high as in Malawi. From 1985 to 1988 they fell by 71 percent.

These data indicate, first, that maize-market policies were inward-looking, and second, that price changes seem to have been instituted to influence domestic production. It would be worthwhile to investigate whether the national price policies have actually contributed to this objective and at what costs. Muir and Blackie (1988) support the view that national policies have contributed to instability in the maize economy.

Producer incentives in the economies differ not only because price levels for individual products differ, but also because price ratios differ (Table 3.17). Since maize is the main staple in the region, all price ratios have been computed relative to the price of maize. Price ratios differ significantly among the countries. Zimbabwe provides significantly higher incentives to producers of coffee, tobacco, and tea than the other countries, even though these exportables are taxed in all four countries. If the other countries were to adopt the price pattern of Zimbabwe, one could expect a significant decline in their maize production. This would necessarily imply a reduction of self-sufficiency in these countries. However, as they would be able to earn more foreign exchange for their main export crops, they would be able to import the needed maize from Zimbabwe. This clearly indicates that trade expansion between these countries is not limited to growth in trade between them, but extends to the rest of the world. Policies would be badly designed if they were to aim at balanced trade only within this group of countries.

Maize production in the individual countries would also be affected if price ratios among the individual grains were to be adjusted. There is greater incentive for sorghum and millet production in Malawi and Zambia than in Zimbabwe, and millet production is significantly depressed in Tanzania. It is not possible to accurately quantify the trade effects of a realignment in prices, because own-price and cross-price elasticities are not available. However, it is known that farmers react to changes in price incentives. Moreover, even small changes in production may give rise to significant changes in trade flows between these countries because all four countries are near self-sufficiency in the main grains apart from wheat and rice.

In a country study on Zambia, Jansen (1988) highlights the effect on agricultural production when internal price ratios are adjusted in line with comparative advantage. Jansen shows that the direct effect of pricing policy in Zambia was to reduce maize output by an average of 27.5 percent a year during 1966-84. The total effect, which included indirect effects from price intervention in other commodities, was even higher, accounting for 58 percent on average for the same period.

The impact of a more flexible trading regime on intraregional trade in maize and maize-related meat can be illustrated with a more recent development in the Zimbabwean maize economy. Prices for yellow and white maize may serve as an example to highlight the issue. The official prices for both types of maize were the same up to 1990. However, production costs for yellow maize are lower than for white maize in Zimbabwe. Yellow maize has higher yields, is more resistant

Table 3.17—Price ratios for selected agricultural products in Malawi, Tanzania, Zambia, and Zimbabwe, 1980-88

Product	1980	1981	1982	1983	1984	1985	1986	1987	1988
Malawi									
Wheat	1.67	1.67	1.96	1.98	1.51	2.00	2.53	2.62	2.30
Rice, paddy	1.55	1.52	0.98	1.14	1.30	1.39	1.42	1.52	1.64
Maize	1.00	1.00	1.00	1.00	1.00	1.00	1.00	1.00	1.00
Sorghum	0.76	0.76	0.45	0.90	0.82	1.12	1.59	0.99	0.84
Groundnuts in shell	5.00	5.00	2.95	4.48	1.75	5.57	2.67	2.44	2.13
Tobacco leaves	6.71	6.47	3.99	6.68	6.79	8.11	7.32	6.98	6.32
Coffee, green	8.94	8.94	7.37	6.63	5.83	6.84	5.34	3.52	3.05
Tea	19.39	21.52	13.51	13.96	13.33	13.52	12.59	11.90	10.34
Tanzania									
Wheat	1.65	1.47	1.43	1.36	1.13	1.14	1.14	1.10	1.13
Rice, paddy	1.75	1.53	1.71	1.82	1.50	1.52	1.52	1.76	1.68
Barley	1.70	1.27	1.44	1.37	1.14	1.14	1.23	1.23	1.26
Maize	1.00	1.00	1.00	1.00	1.00	1.00	1.00	1.00	1.00
Sorghum	1.00	0.67	0.91	0.91	0.75	0.76	0.76	0.73	0.71
Groundnuts in shell	4.20	3.20	3.31	3.64	3.20	3.41	3.41	3.28	3.10
Tobacco leaves	10.50	8.40	10.29	8.18	6.30	7.22	7.82	7.68	7.25
Coffee, green	9.00	6.00	6.86	7.64	5.88	5.37	8.06	7.43	7.62
Tea	1.50	1.00	1.14	1.27	1.03	0.94	1.21	1.21	1.26
Millet	1.50	1.00	0.86	0.91	0.75	0.76	0.76	0.73	0.69
Soybeans	2.25	1.50	1.71	2.05	1.69	1.79	1.79	1.73	1.64
Zambia									
Wheat	1.71	1.93	2.00	1.96	1.74	1.59	1.57	1.90	2.99
Rice, paddy	1.73	1.55	1.75	2.19	1.63	1.41	1.14	1.60	1.96
Maize	1.00	1.00	1.00	1.00	1.00	1.00	1.00	1.00	1.00
Sorghum	0.52	0.67	0.56	0.88	0.76	0.95	0.78	1.26	1.19
Groundnuts in shell	2.25	2.39	2.26	2.27	1.71	2.44	1.80	2.09	3.44
Tobacco leaves	12.08	11.00	13.48	13.30	10.29	10.95	8.38	8.03	8.06
Coffee, green	14.65	13.35	16.35	16.65	14.41	14.41	8.63	9.42	9.34
Tea	14.65	13.35	0.00	0.00	0.00	0.00	0.00	0.00	0.00
Millet	0.52	0.45	0.38	1.62	1.21	1.34	1.02	1.57	2.51
Soybeans	2.73	2.69	2.64	2.48	2.14	2.15	2.04	2.53	3.42
Zimbabwe									
Wheat	1.79	1.45	1.56	1.83	1.78	1.59	1.67	1.84	1.81
Rice, paddy	1.81	1.73	1.83	2.08	1.96	1.68	1.85	2.04	2.03
Maize	1.00	1.00	1.00	1.00	1.00	1.00	1.00	1.00	1.00
Sorghum	1.10	0.90	0.88	0.91	0.99	0.96	0.97	1.56	1.53
Groundnuts in shell	2.91	2.23	2.33	2.32	2.18	2.55	2.57	2.97	2.85
Tobacco leaves	8.93	15.45	13.94	15.71	14.76	14.99	17.61	12.20	11.40
Coffee, green	24.72	12.71	14.48	17.89	21.34	21.61	30.78	21.00	21.24
Tea	11.43	10.90	12.17	13.75	13.36	11.79	13.43	15.08	15.03
Millet	0.84	0.85	1.00	1.17	1.19	1.09	1.29	1.53	1.50
Soybeans	1.80	1.42	1.66	2.14	2.04	1.78	1.89	2.14	2.12

Source: Food and Agriculture Organization of the United Nations, "Producer Prices Tape," FAO, Rome, 1989.

Note: All prices are relative to the price of maize.

to drought, is less prone to theft, has an even higher feeding value than white maize, but is less preferred for human consumption. Given the present system of incentives, there is a strong incentive to produce yellow maize. Hence, the yellow maize economy has shown a remarkable dynamic. Yellow maize production increased by more than 1,000 percent from 1985 to 1990 and accounts for 38.5 percent of the area cultivated with maize. At the same time, the poultry industry has grown fast over recent years. Maize sold by the marketing board to stockfeeders increased by 372 percent from 1988 to 1990, and that sold to poultry farmers by 272 percent.

Undifferentiated prices between white and yellow maize affect the market for white maize, and thus its exportable surplus, in two ways. First, white maize production is lower than it would be if yellow maize prices were lower, and second, undifferentiated prices do not guarantee that only yellow maize is fed to animals. Consequently, white maize is less available for human consumption. Free internal trade for yellow maize at lower prices than for white maize would have one additional positive effect. Producers of yellow maize would receive incentives to trade their produce directly to stockfeeders. They could trade using their own vehicle during the off-seasons, reducing the drain on the private and the grain marketing board's transport systems. The drain on the storage economy would also be reduced. The benefit of lower feed prices would further boost the livestock sector, especially the poultry sector, and would increase exportable surpluses of white maize and meat.

Market integration would also lead to adjustment in consumer prices and could help to open up national markets for trade. Consumer prices for maize flour differ widely among the four countries (Table 3.18). Prices were lowest in Zambia in 1986; only three months' salary was needed to buy 1 metric ton of maize flour.

Table 3.18—Consumer price for maize flour and purchasing power for selected countries of the Southern African Development Coordination Conference, 1986

Price/Wage	Malawi	Tanzania	Zambia	Zimbabwe
Price per metric ton (US$)	97	n.a.	88	140
Average minimum monthly wage (U$)	22	31	29	36
Wage months needed for 1 metric ton of maize flour	5	n.a.	3	4

Source: Preferential Trade Area, *Report of the Task Force Study on Subregional Food Marketing* (Lusaka: PTA, 1988).

Note: n.a. means not available.

Adjusting the price for maize flour would decrease consumption in low-price countries like Zambia, and would increase consumption in high-price countries like Tanzania. Consequently, a new pattern of deficits and surpluses would emerge.

It should be noted that present differences in consumer and producer prices among the countries might be quite costly for governments, as they create an incentive for parallel markets and illegal trade across borders.

Barriers to Intraregional Trade and Policies to Remove Them

There seems to be ample evidence that a significant potential for intraregional trade exists. One may wonder why countries have not yet taken advantage of more intensive division of labor. It will be argued in the following section that the main barriers to trade have been specific elements of national policies and, therefore, changes in policies are needed to promote intraregional trade. However, such policy changes will most likely be accepted only if policymakers are convinced that removing the policy barriers to intraregional trade will, in balance, positively contribute to the achievement of their national objective. Hence, it will be argued implicitly that past and present policies did not contribute to the achievement of overall objectives as much as they could have, partly because they were hindering intraregional trade.

The External Trading Regime. The trading regime of all four countries under investigation is very much dominated by the shortage of foreign exchange, yet the foreign-exchange problem is at the heart of the trade problem. Overvalued currencies tax exports and reduce the potential export volume. Moreover, imports from neighboring countries are smaller because countries prefer to export to hard-currency countries. A shortage of foreign exchange normally leads to administrative rationing measures; license systems and foreign-exchange allocation systems are the by-product of overvalued currencies. As a result, the public sector employs resources that could be used more productively in the private sector. In addition, the private sector, especially the trading sector, becomes less efficient because of the myriad of regulations. Foreign-exchange regulations contribute to a significant increase in transaction costs (Rusike 1989). Governments nevertheless continue to overvalue their currencies in the belief that they contribute to food security and positive income distribution. However, evidence supports the opposite.

When a country's currency is overvalued, the international division of labor is not fully absorbed by its economy, and the use of domestic resources is less efficient than it would be otherwise. Moreover, countries with overvalued currencies attract less capital inflow and stimulate capital outflow; consequently, resources employed in the internal economy are reduced, overall income is lower, and food security on the national level is threatened. These predictions derived from economic theory are largely supported by the experience of those countries that moved to a more open trading regime. A World Bank team, Papageorgiou, Choski, and Michaely (1990), has recently published the findings of the experience of 19 countries that have undergone phases of liberalization. In general, well-designed liberalization policies improved the countries' economic performance. However, they found that the program should begin with a substantial real depreciation of the currency.

Such a move to liberalization would also promote intraregional trade because it would allow countries to adjust their production and consumption patterns in accordance with comparative advantage. Moreover, countries that had been reluctant to export to neighboring countries, due to preference for hard currencies, could relax their trading policies.

Most SADCC countries try to ease the foreign-exchange constraint by specific export-promotion schemes. However, such a strategy is more likely to be counterproductive in achieving market integration. Promotion of intraregional trade should not only promote exports but also enhance imports. Export-promotion schemes result in implicit multiple exchange-rate systems that further distort the national economies. Likewise, promoting schemes of intraregional trade without solving the foreign-exchange-rate problem is probably doomed to fail. Regional integration cannot be a substitute for opening up the economies worldwide. Instead, worldwide liberalization will most likely promote intraregional trade. When individual countries earn more foreign exchange by exporting to the world market, they earn the purchasing power to import from their neighboring countries, which may offer better terms than countries outside the region.

Furthermore, past and present policies are often supposed to have positive distribution of income effects and thus contribute to food security on the household level. However, this perception has to be challenged based on the findings of the experience of those countries that instituted liberalization policies. Papageorgiou, Choski, and Michaely (1990) found no empirical evidence to support this widely held view. Specific information on the four countries under study suggests that liberalization would most likely improve distribution of income.

1. As shown above, the present system results in a heavy tax on agriculture; therefore, liberalization would increase the agricultural sector's income. Since a large share of the poor's expenditure comes from agriculture, this may improve their food-security situation.

2. Moreover, devaluation of the currency would promote agricultural exports and would most likely have a positive effect on employment as export crops are labor-intensive. Reallocating acreage from cereal production to cotton, tea, coffee, and tobacco would at least increase labor input by 50 percent per cultivated hectare (Msomba 1990).

3. Finally, governments may believe that they are able to enforce the present system of trade regulations. However, if prices between countries vary to the extent presented above, and if overvalued currencies affect commodities differently across borders, there are built-in incentives to circumvent the regulations. Illegal transborder trade is only one of the many consequences. Even more important and most likely are over- and under-invoicing of official foreign-trade transactions and transactions on the parallel market for foreign exchange. Yeats (1989) found that African trade data are not consistent. Import data do not match with export data of other countries. This indicates that African governments are not reliably informed on their trade situation. Moreover, it proves that a significant part of trade may go through unofficial channels. The inability to control the external trading sector can lead to high costs for the economy. It may be that directly or indirectly subsidized products in one country may be shipped to neighboring countries; for instance, maize meal from Zambia to Zaire, where it fetches a higher market price. Hence, the exporting country is losing. What it receives in return is of less value than what it has delivered.

However, it is not only the exchange-rate problem that hinders intraregional trade. It is also the present external trading regime in some of those elements that are not connected to the exchange-rate problem.

Tariff and nontariff barriers to trade have been identified as the most formidable impediment to the expansion of trade among developing countries (FAO 1987). A General Agreement on Tariffs and Trade (GATT) study states that tariffs on agricultural products are generally higher than tariffs on manufactured products and that "processed agricultural products generally face higher tariff and nontariff restrictions than primary commodities in most developing countries" (GATT, n.d.). This is of special relevance for the four countries under investigation. Some of them have set up food-processing companies that are run at high costs because the narrowness of the internal market does not allow them to use their capacities fully and to exploit economies of scale.

Internal Policies. Internal policies are allegedly instituted to improve food security, but the distortion of the internal production and consumption pattern by setting wrong price signals and the strong regulation of internal trade may affect intraregional trade and food security in a way unintended by policymakers.

National price policies. It is obvious that price ratios that are out of line among the neighboring countries affect domestic production and consumption. Traditionally, policymakers in these countries aimed at achieving food security on the national level and often confused food security with food self-sufficiency. Such policies decrease import demand and thus conflict with other countries' export chances. Consequently, trade is reduced.

Fortunately, food security is now viewed in a much broader sense. In August 1990, the SADCC regional program redefined food security in a way that extends the limited concept of national food self-sufficiency to regional self-sufficiency. Thus, policymakers have accepted that regional trade can play a major role in achieving overall food security without sacrificing national objectives. However, it has to be noted that even this extended definition of food security will not allow the intraregional trade potential to be exploited to its full extent. Individual countries and the region as a whole might well be better off if they exported nonstaple foods and other agricultural products to outside the region and imported the needed food. Certainly, food security can be achieved only if a sufficient amount of food is available. However, food can be available in a country even if it is not produced there. A strategy that leads to more imports of food—an increase in food dependency—may improve food security. Under this scenario the country uses fewer domestic resources to produce food and more to produce exportables and to generate foreign exchange that allows it to import food.

Internal marketing of agricultural products. The internal marketing system impedes intraregional trade in two ways: by allowing government to set prices, and by restricting private trading.

Governments prefer to set uniform regional and seasonal prices for agricultural raw products as well as for processed agricultural products. The policy of panterritorial pricing does not allow specialization of the individual regions in consideration of their comparative advantage and consequently hinders the most efficient use of available resources.

Moreover, this policy places a special drain on the internal transport sector, as the agricultural produce has to be bought in by the public marketing agents soon after the harvest and must be transported to the centers for milling or stockpiling. Later in the year the processed products have to be shipped to the countryside for consumption. Giving up panterritorial and panseasonal pricing would place a smaller seasonal drain on the public transport sector and on public storage, as some of the produce would be stored on the farms. Transport costs would also be lower because the transport system would become more and more the responsibility of the private sector, which uses traditional and cheaper transport modes (at present, official trucks are often empty on the return trip).

It is argued that the present policies are more efficient in achieving food security. However, new research findings support the opposite. Jayne, Chigume, and Hedden-Dunkhorst (1990) argue that the present system in Zimbabwe is based on the notion that households in rural areas—even those that are poor—produce a surplus of grain above annual consumption and that panterritorial and panseasonal prices can be enforced at any location at any time within a given year. However, the empirical evidence shows that there are many grain deficit households in rural areas and that they have to buy maize meal at a significantly higher price. Hence, their food-security status is negatively affected by the present system of pricing. Concerning the enforcement of panseasonal and panterritorial pricing, there is similar evidence from the other three countries. Enforcement is not possible—at least not in rural areas—and price swings are most likely larger than they would be in a more liberal market with much lower transaction costs for private traders.

The present marketing system with official price-setting and the institution of parastatals necessarily implies that governments have to subsidize traders. This system is maintained at a very high cost. First, parastatals usually run a high deficit due to the government's price-setting, and they tend to be inefficient because they lack competitors. When private trade exists, the absence of marketing institutions, like price-information schemes, and the high-risk nature of the illegal private trading impose higher transaction costs than in other more liberal markets. The present organization of the internal marketing system is an obstacle to intraregional trade because internal trade flows are determined by government restrictions and thus are not permitted to cross national borders. Moreover, experience from other countries indicates that parastatals are not sufficiently flexible to respond to changes in market conditions in neighboring countries.

It is worthwhile to note that even the SADCC council accepted that restrictions in intraregional trade impeded food security. "By constraining intraregional trade in staple food commodities during 1988/89, these problems lessened the degree of internal food security which the region could have otherwise enjoyed. Unless those constraints are urgently resolved, they will have a similar negative impact on regional food security in the current marketing year" (SADCC 1990).

High transaction cost. Governmental interference in the marketing system is not the only contributor to high transaction costs in intraregional trade. Transaction costs are defined as those costs that have to be incurred in transferring commodities from producers to consumers. Thus, transaction costs result from transforming products in space and time.

The existing market infrastructure inside each country connects rural areas to a major city but not to each other. The lack of direct economic and physical links between the rural areas increases the transaction costs for internal trade between them. This holds even more for transborder trade. Therefore, the improvement of the transport system has to be seen as a major step in promoting intraregional trade.

Limited market information, restricted travel, and the absence of any reliable means of legally enforcing the commitments of contract partners are the lot of private traders when governments institute marketing boards. This situation contributes to high transaction costs.

It has been shown that market integration among countries has to follow market integration within the individual countries. Yet SADCC has inherited the present marketing system that favors trading with overseas countries, and therefore they need to provide the necessary trading infrastructure for intraregional trade promotion.

The Role of Food Aid. The basic idea presented in previous sections—that shortfalls in some countries could be compensated for by surpluses in other countries—cannot work if food aid is delivered from outside the region. Moreover, the intraregional trade potential is occasionally eroded because of ill-timed food-aid deliveries.

Zimbabwe enjoyed a bumper crop in 1985 with 3.56 million tons, about 124 percent higher than in 1984. Malawi's 1985 harvest was about the same as in 1984, and Zambia harvested 29 percent more in 1985 than in 1984. Altogether, cereal production of the three countries would have allowed them to increase their 1981-83 cereal consumption by 83 percent. Thus their surplus was much greater than their neighbor, cereal-deficit Mozambique, could have used. In the same year, 1985, Zimbabwe received food aid of 131,300 tons, an amount much higher than ever received before or after. Similarly, Malawi, Mozambique, and Zambia accepted significant amounts of food aid in 1985. Their total aid amounted to about 15 percent of cereal consumption in 1981-83 in spite of having produced internally an excess of 67 percent over cereal consumption in the same period. It is true that food aid did help to close the actual cereal gap in two out of the four neighboring countries. However, food aid even increased the surplus in two countries and contributed significantly to the regional surplus. The consequences are of special concern for surplus-producing countries. Faced with low export parity prices for exports to overseas, surplus countries had to stock up; Zimbabwe's maize stocks eventually amounted to more than annual consumption.

Although, the situation described above is not at all a necessary outcome of food aid, it is not an exception and has become the rule for the last five years. It indicates how food aid—if badly timed—can have negative effects on intraregional trade.

The situation for the main cereals, maize, wheat, and rice, for the years 1985-89 is presented in Figures 3.2 to 3.4. A comparison of national surpluses for the individual grains and food aid clearly shows that individual countries received food aid even if they had a surplus, and in all five years the region as a whole received more food aid than was needed to compensate for the regional deficit in maize, and occasionally even for rice.

CONCLUSION AND STRATEGIES TO PROMOTE INTRAREGIONAL TRADE

Major policy changes are needed if the trade potential is to be exploited. A review of worldwide integration or cooperation schemes conveys one clear message: regional integration cannot be a substitute for appropriate domestic economic policies (Langhammer and Hiemenz 1989). The disappointing progress or even the failure of integration schemes is not due to a lack of integration potential. "With wrong exchange rates, high barriers to trade, misleading investment incentives, and a high degree of governmental participation in economic activities, patterns of trade are distorted in such a way that RIDC (Regional Integration of Developing Countries, U.K.) does not offer many benefits to those presently gaining from the domestic economic regime" (Langhammer and Hiemenz 1989). Therefore they oppose regional economic liberalization as much as they oppose opening up toward world markets. Hence, the lesson that can be drawn from the

Figure 3.2—Food balances and food aid: maize, 1985-89

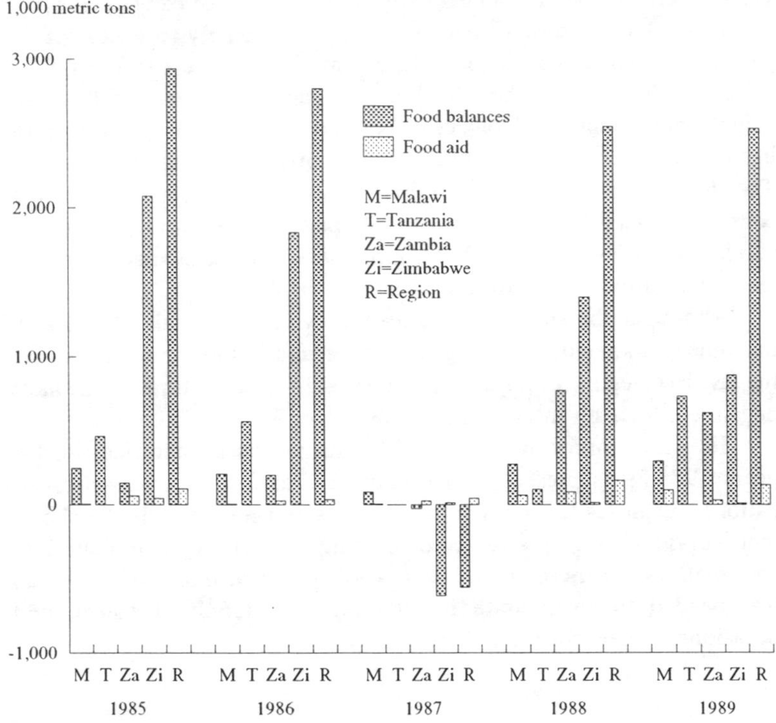

Source: Southern African Development Coordination Conference, *Food Security Bulletin* (Harare: SADCC, various years).

past experience is that internal liberalization is the only starting point for a program of regional integration.

Governments can hardly be expected to institute drastic policy changes in one stroke. However, there might be some strategies, through a combination of unilateral and concerted actions, that could help capture some of the trade benefits in a transition period even before the total package of needed policy changes is instituted.

Unilateral Actions

Unilateral actions should aim to adjust the internal production and consumption patterns to international prices—internal prices have to be adjusted and the domestic economy has to be deregulated. Finally,

Figure 3.3—Food balances and food aid: wheat, 1985-89

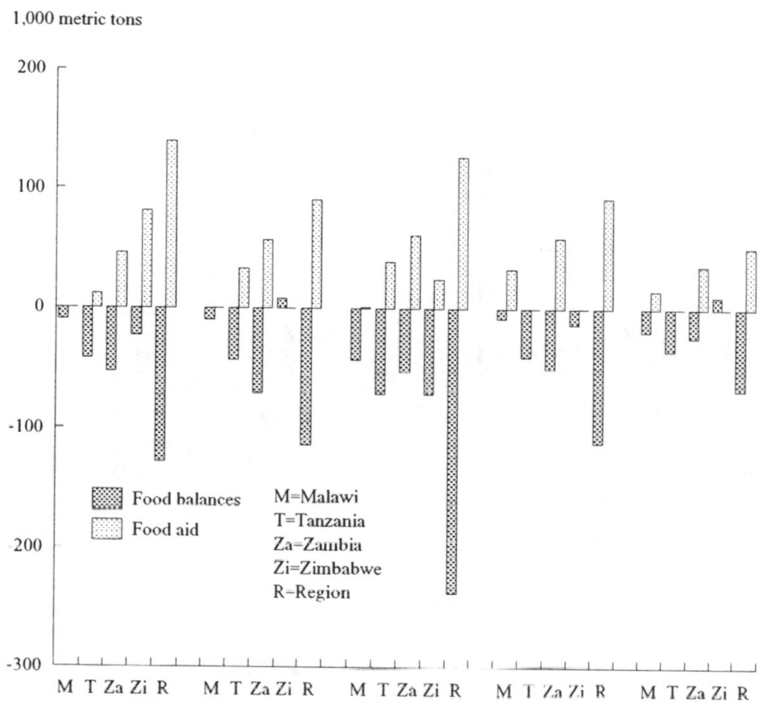

Source: Southern African Development Coordination Conference, *Food Security Bulletin* (Harare: SADCC, various years).

policies should encourage the privatization of markets and improve the market infrastructure in order to facilitate the integration of the domestic economy in a larger region.

Concerted Actions

In some instances, concerted actions by a group of countries to promote trade may be more efficient than unilateral actions.

1. There may be national public goods that could be produced more cheaply on a regional level and also regional public goods that could be produced more cheaply if countries cooperated.

2. Countries as a group could aim at avoiding duplication of investment in food-processing industries when national markets are too

Figure 3.4—Food balances and food aid: rice, 1985-89

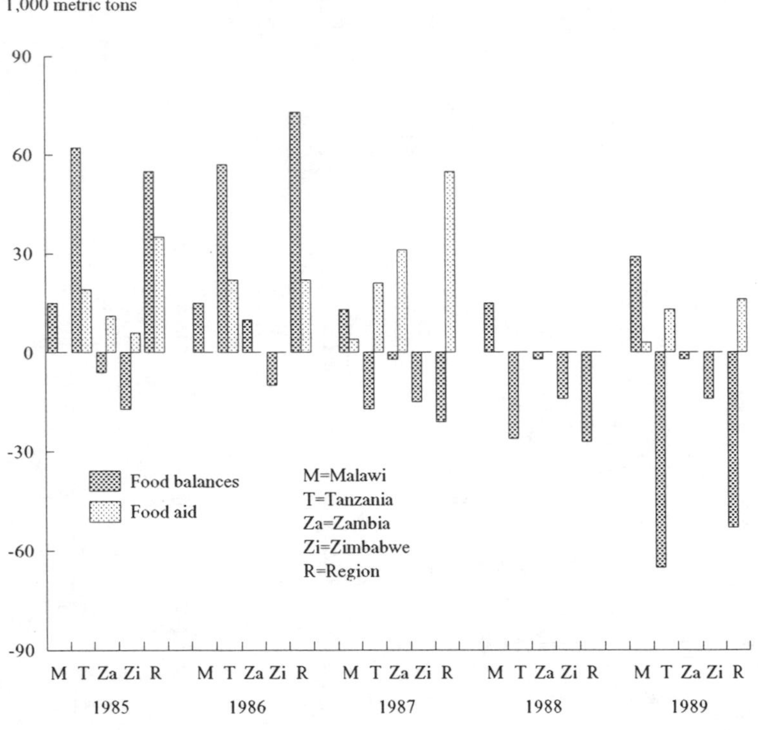

Source: Southern African Development Coordination Conference, *Food Security Bulletin* (Harare: SADCC, various years).

narrow to take full advantage of economies of scale. This would work toward a more efficient way of achieving food security.

3. Transaction cost, a major obstacle to intraregional trade, may be lowered through such actions as trading arrangements that guarantee secure access to markets; regional marketing intelligence that provides market information for private traders; improvement of the marketing infrastructure, including the introduction of a grading system and its surveillance; and institutions that support cross-border trading activities.

4. First, food aid could be used to promote trade, and trade could be used to make food aid more efficient for the region as a whole. Food aid could be given in the form of "triangular" purchases involving a donor and either two aid recipients or a commercial importer and one food aid

recipient. Actually, triangular purchases have increased in recent years, and most grain trade in the region is based on this type of transaction, but the past record indicates that triangular transactions could be improved. The type of food aid received must match the type of grain in deficit in the region. Hence, food aid actually provided by donors should not be based on a specific country's production shortfall in one specific type of grain, but rather on a shortfall on the regional level.

Second, the timing of this transaction within a given year could be improved. The Food Security Unit of SADCC has set up an efficient early warning system that provides information on predicted production early in the year. The countries could also make more efficient use of their grain marketing boards. Figures 3.5 and 3.6 show the monthly purchases of the grain marketing board. Most of the grain is bought within a short period of time. Hence, the governments in the individual countries know quite early in the year what the actual harvest will be. They are therefore able to contract purchases early in the year, preferably before the harvest in the neighboring countries has been bought in. Donors could be approached early in the year if food aid is needed, but food aid should not be used if there is sufficient food on the regional level even though there may be insufficient food on the national level. A specific problem of intraregional trade concerns pricing of traded commodities. National prices and official exchange rates are not acceptable guidelines because they do not reflect national shadow prices of the exporting country or the importing country. Possibly these countries could draw on the experience of some other regional trading blocs, like that of the East bloc countries, which have agreed to trade on the basis of international prices and to pay in hard currencies. In order to shorten the negotiating period, a decision on which international prices to use has to be made in advance in the case where actual deliveries could materialize.

5. It was shown earlier in this chapter that there is a potential for redirecting trade flows. Countries do not trade some products with each other because they are not aware of the market situation in the neighboring countries or because they prefer to export to hard-currency countries. The first deterrent could be overcome by providing better market information, the second by agreeing to trade in hard currencies at international prices. However, such an agreement would allow exploitation of only a small percentage of the trade potential. It should be restricted to those products for which world market prices can be easily specified and where trade can be controlled. Hence, it may open up a trade potential for fairly homogeneous commodities that are traded by parastatals.

Figure 3.5—Maize purchases by marketing authorities, Malawi, 1987 and 1988

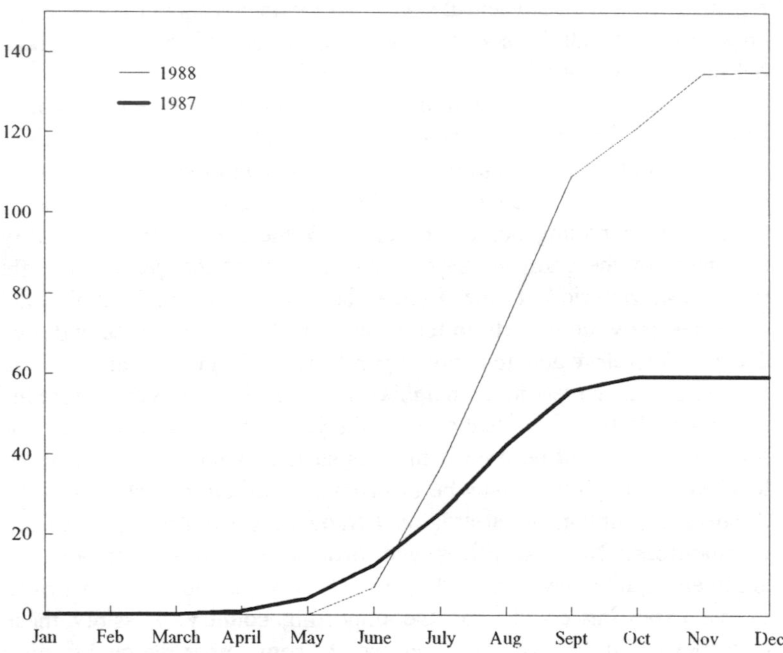

Source: Southern African Development Coordination Conference, *Food Security Bulletin* (Harare: SADCC, 1989).

Such a system could easily be introduced, especially on the grain market. As was indicated earlier, governments know early in the year whether they have an exportable surplus or need to import. Agreements for trade should be made as early as possible within the year in order to minimize transport costs. Transports from the surplus areas to the silos in the same country can be quite costly; instead, the flow of products would go directly from the surplus area in one country to the nearest deficit area in a neighboring country.

Figure 3.6—Maize purchases by marketing authorities, Zimbabwe, 1987 and 1988

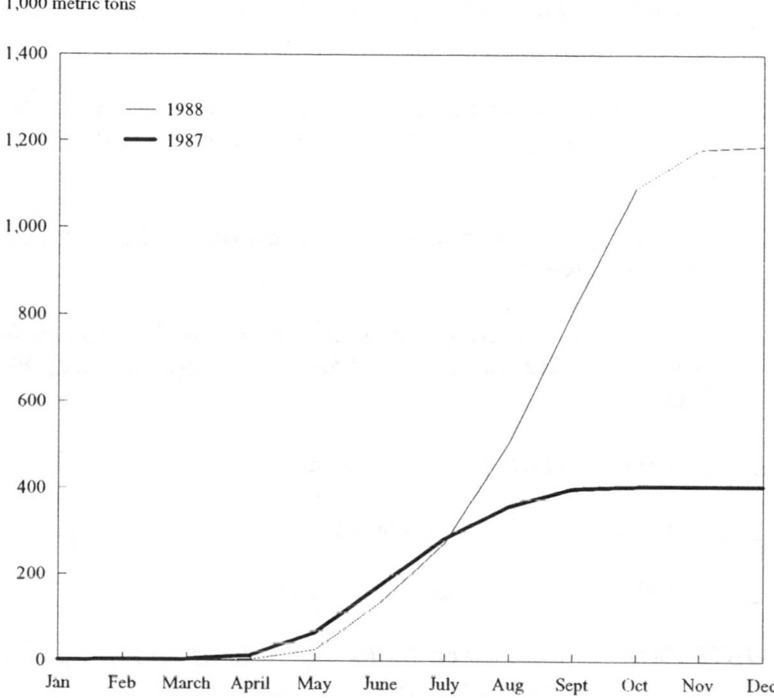

Source: Southern African Development Coordination Conference, *Food Security Bulletin* (Harare: SADCC, 1989).

REFERENCES

Finger, J. M., and M. E. Kreinin. 1979. A measure of "export similarity" and its possible uses. *The Economic Journal* 89 (356): 905-912.

FAO (Food and Agriculture Organization of the United Nations). 1984. *Agroclimatological data for Africa*. Plant Production and Protection Series. Rome.

_____. 1986. *African agriculture, the next 25 years: Atlas of African agriculture*. Rome.

_____. 1987. *Economic cooperation among developing countries in agricultural trade*. Economic and Social Development Paper 70. Rome.

_____. 1989a. Producer prices tape. Rome.

_____. 1989b. Production yearbook tape, 1988. Rome.

_____. 1990. Food balance sheets tape. Rome.

GATT (General Agreement on Tariffs and Trade). N.d. *Expansion of trade among developing countries*. LDCITS/1-50, Technical Studies 1978-79. Geneva.

International Food Policy Research Institute. Inter-LDC trade data base. International Food Policy Research Institute, Washington, D.C., 1987.

International Monetary Fund. 1990. *International finance statistics*. Washington, D.C.: IMF.

Jansen, D. 1988. *Trade, exchange rate, and agricultural pricing policies in Zambia*. A World Bank Comparative Study in The Political Economy of Agricultural Pricing Policy series. Washington, D.C.: World Bank.

Jayne, T. S., S. Chigume, and B. Hedden-Dunkhorst. 1990. Unravelling Zimbabwe's food security paradox: Implications for grain marketing reform. Paper presented at the First National Consultative Workshop on Integrating Food, Nutrition, and Agricultural Policy, 15-18 July, Juliasdale, Zimbabwe.

Koester, U. 1990. Agricultural trade among Malawi, Tanzania, Zambia, and Zimbabwe. Paper prepared for the World Bank, Agricultural Division, Southern Africa Department. International Food Policy Research Institute, Washington, D.C. Mimeo.

_____. 1986. *Regional cooperation to improve food security in southern and eastern African countries*. Research Report 53. Washington, D.C.: International Food Policy Research Institute.

_____. 1984. Regional cooperation among developing countries to improve food security. *Quarterly Journal of International Agriculture* 23 (April): 99-114.

Langhammer, R. J., and U. Hiemenz. 1989. Regional integration among developing countries: Survey of past performance and agenda for future policy action. The Kiel Institute of World Economics, Kiel, Germany. Mimeo.

Lipton, M. 1988. Regional trade and food security in southern Africa. In *Poverty, policy and food security in southern Africa*, ed. C. Bryant, Boulder, Colo., U.S.A., 93-121. Lynne Rienner and London: Mansell.

Msomba, N. T. 1990. Beurteilung der Getreidemarktpolitik Tansanias. Diplomarbeit, University of Kiel, Kiel, Germany.

Muir, K., and M. Blackie. 1988. *Maize price cycles in southern and eastern Africa*. Working Paper AEE 6/88. Harare: University of Zimbabwe, Department of Agricultural Economics and Extension.

Preferential Trade Area. 1988. *Report of the task force study on subregional food marketing*. Lusaka: PTA.

Papageorgiou, D., A. M. Choski, and M. Michaely. 1990. *Liberalizing foreign trade in developing countries. The lessons of experience*. Washington, D.C.: World Bank.

Rusike, J. 1989. *Trader perceptions of constraints on expanding agricultural input trade among selected SADCC countries*. Working Paper AEE 5/89. Harare: University of Zimbabwe, Department of Agricultural Economics and Extension.

SADCC (Southern African Development Coordination Conference). 1990. *Food, agriculture and natural resources*. Lusaka: SADCC.

_____. Regional Food Security Programme. 1984. Regional food reserve. Main report of a prefeasibility study prepared by Techno-synesis, Zimbabwe.

_____. Various years. *Food Security Bulletin*. Zimbabwe, Harare.

United Nations Conference on Trade and Development. 1989. Export value tape. Geneva.

Yeats, A. J. 1989. *On the accuracy of economic observations. Do Sub-Saharan trade statistics mean anything?* International Economics Department Working Papers. Washington, D.C.: World Bank.

4

Agricultural Incentives and International Competitiveness in Four African Countries: Government Interventions and Exogenous Shocks

Jorge Quiroz and Alberto Valdés

The influence of sector-specific and economywide policies in the evolution of agricultural prices in Malawi, Tanzania, Zambia, and Zimbabwe and the ways these policies may influence the final outcome of trade-liberalization efforts and economic-integration schemes are analyzed in this chapter. In general, government intervention may affect relative agricultural prices through direct or indirect effects (Krueger, Schiff, and Valdés 1988). A direct effect may appear when the price of a given product is influenced by the imposition of trade barriers specific to the product or by the action of a parastatal agency. Indirect or economywide effects typically result from the impact of macroeconomic policies on the real exchange rate and thereby on the relative price between tradables and nontradables. Moreover, the actual evolution of incentives is also influenced by exogenous factors, which in turn may influence domestic policies.

The role of government interventions in the determination of relative prices must be addressed in any proposal that considers, for example, the potential benefits of further economic integration. A related issue concerns the adjustment of relative agricultural prices that countries should undertake if the removal of some key distortions in the economy is to be pursued.

This chapter deals with three questions: (1) What was the level and evolution of the structure of agricultural incentives in these four countries during 1980-87? (2) What part of the evolution of agricultural prices can be attributed to exogenous factors (primarily international prices) and what part is probably the result of domestic policies (government interventions)? (3) What adjustment in agricultural prices can be expected if relative price distortions at the macro and micro level are to be removed?

The answers to these questions carry immediate implications for the analysis of trade flows between the countries under study. From (2) may be seen whether domestic policies have either induced or inhibited trade among the countries in specific products. From (3) one may infer what the most likely trade adjustment among countries would be if these countries pursued some trade reforms and macroeconomic corrections simultaneously.

The basic framework for analyzing the evolution of agricultural prices is presented in the next section. The change in agricultural relative prices is decomposed into three parts: the change in the international price, the change in direct price intervention (nominal protection), and the change in the real exchange rate. Clearly, changes in international prices are exogenous to the countries under study, while changes in the nominal protection rate are the result of government intervention. On the other hand, changes in the real exchange rate depend on both exogenous and policy-related variables. Next, the main determinants of the real exchange rate are discussed along with the econometric results of a simple real exchange rate model aimed at distinguishing between exogenous and policy-related determinants of the real exchange rate. All of these results are then used to decompose the evolution of agricultural prices into exogenous and policy-related components. In another section, the results of the model for the real exchange rate are used again, together with some estimates of nominal rates of protection, to determine the possible adjustment in relative agricultural prices that would take place if some of the trade distortions were removed. The last section summarizes the main conclusions.

A DECOMPOSITION OF AGRICULTURAL PRICES

The objective of this section is to decompose the evolution of agricultural prices into three main determinants: changes in international prices, changes in the nominal protection rate, and changes in the real exchange rate. To this end, denote by P_{it} the nominal price of agricultural good i at time t, measured in domestic currency, and denote by CPI_t the level of the consumer price index at time t. The consideration here is the evolution of relative agricultural price incentives over time, namely, the evolution of

$$p_{it} = \frac{P_{it}}{CPI_t}. \tag{1}$$

Denote by P_{it}^* the corresponding border price the country faces (c.i.f. for importables and f.o.b. for exportables), measured in foreign currency (U.S. dollars). Let E_t be the level of the nominal exchange rate (measured in units of domestic currency per U.S. dollar) at time t. By definition,

$$P_{it} = P_{it}^* E_t \, (1 + \gamma_{it})(1 + t_{it}), \tag{2}$$

where γ_{it} is meant to be a "markup" factor including transport costs and competitive profit margins to make the border price comparable with the domestic price, and t_{it}, the residual after the markup, is meant to be the nominal protection rate.

It must be stressed that the markup factor used here corresponds to a hypothetical transport cost and competitive profit margin consistent with a counterfactual scenario where trade would be carried out only by private agents. As a consequence, the nominal protection rate considered here does not conceptually coincide with the "explicit" nominal export or import tariffs that the government may impose. Due to the widespread involvement of government agencies in the trade of agricultural products in these countries, the actual profit margin may very well be above the competitive profit margin, thus implying an implicit tax rate superior to the explicit import or export tariff. Also, and perhaps more important, an implicit import or export tariff may result from inefficiencies arising from the operations of the different parastatals involved in agricultural trade.

Denoting by CPI_t^* the general level of the foreign prices at time t (U.S. CPI) and using equation (1), it is easily seen that

$$p_{it} = \left[\frac{P_{it}}{P_{it}^* E_t} \right] \left[\frac{P_{it}^*}{CPI_t^*} \right] \left[\frac{CPI_t^* E_t}{CPI_t} \right], \tag{3}$$

and using equations (2) and (3) above,

$$p_{it} = (1 + \gamma_{it})(1 + t_{it}) p_{it}^* \; RERt, \tag{4}$$

where RER denotes the real exchange rate, defined as the ratio of international and domestic prices, and p^* has the obvious interpretation of the foreign counterpart of equation (1). The RER so defined is in agreement with most empirical definitions used in the literature and can be taken as a proxy for the theoretical counterpart of the relative price of tradables against nontradables.[1]

In practice, $(1 + \gamma_{it})(1 + t_{it})$ can be obtained directly by simply comparing border and domestic prices, but it is very difficult to disentangle the above expression into its primary components without fully analyzing the complete market structure of the good in question. However, for the analysis of the evolution of agricultural prices, it may be reasonable to assume that the markup factor γ remains constant over time, especially if it is recalled that, conceptually, γ is made up of reference competitive costs and profit margins and not of the actual costs and profits. Adopting that assumption then, and taking natural logs on both sides of equation (4), and taking first-order differences, yields

$$\Delta \bar{p}_{it} = \Delta \bar{p}_{it}^{*} + \Delta \overline{RER}_{t} + \Delta \overline{(1 + t_{it})}, \tag{5}$$

where the operator Δ denotes first differences, $\Delta x_t = x_t - x_{t-1}$ and the bar denotes natural logs, $\bar{x} = \log(x)$.

Equation (5) decomposes the evolution of domestic relative agricultural prices into three main factors: the evolution of international relative prices (p^*), the evolution of the real exchange rate (*RER*), and the evolution of direct government intervention or nominal protection (1 + t). The evolution of international prices is mostly exogenously determined from the country's point of view. On the other hand, the evolution of the nominal protection rate is clearly a policy variable. Finally, the evolution of the real exchange rate is determined by both exogenous and policy-related factors.

Tables 4.1 and 4.2 present the evolution of the relative price of agricultural goods, decomposed according to equation (5). To avoid carrying the analysis over a huge number of tables, the decomposition was calculated for aggregate goods only, namely, agricultural exportables and importables for each country. The relative price in each category (exportable, importable) was calculated as a divisia price index with the weighing shares corresponding to the shares in total exports or imports. The divisia price index exhibits a number of desirable properties described in the literature (Barnett 1982). In addition to those properties, and for the case at hand, it yields a well-defined aggregate nominal protection, whose evolution is also presented in Tables 4.1 and 4.2.

---

[1] For an analysis of a possible bias with the use of such proxy for Zimbabwe see Masters 1991.

Table 4.1—Variation in log of relative price of agricultural exportables decomposed by source, 1980-87

Country	Period	Change in Domestic Price (1)	Change in Foreign Price (2)	Real Exchange Rate (3)	Direct Protection (4)
Malawi	1980-82	-15.8	15.5	22.1	-53.5
	1982-85	6.0	-71.7	18.3	59.4
	1985-87	-19.7	49.0	...	63.7
	1980-87	-29.5	-7.2	35.4	-57.8
Tanzania	1980-82	-21.2	-42.9	-20.0	41.7
	1982-85	16.1	-29.3	-9.8	55.1
	1985-87	8.6	21.4	81.5	-94.3
	1980-87	3.5	-50.8	51.7	2.6
Zambia	1980-82	7.4	6.1	7.2	-6.0
	1982-85	69.0	-71.5	64.7	-62.2
	1985-87	26.9	63.3	39.0	-75.4
	1980-87	-34.6	-2.1	110.9	-143.5
Zimbabwe	1980-82	12.2	-14.4	10.0	16.5
	1982-85	7.6	-12.7	39.1	18.8
	1985-87	-39.8	2.5	-16.3	-25.9
	1980-87	-20.0	-24.5	32.8	-28.2
All countries	1980-82	-4.4	-8.9	4.8	-0.3
	1982-85	-9.8	-46.3	28.1	8.4
	1985-87	-6.0	34.0	24.8	-64.8
	1980-87	-20.2	-21.1	57.7	-56.7

Notes: Numbers correspond to 100 times the change in the logs; therefore, figures can be approximately interpreted as percents. The domestic price is decomposed so that (1) = (2) + (3) + (4).

Due to lack of information on producer prices, the decomposition of agricultural prices was calculated only up to 1987. For Tanzania, only the results during the first one and a half years of the reform, which started in mid-1986, appear in Tables 4.1 and 4.2.

Table 4.1 reveals a number of interesting regularities between the different determinants of the price of agricultural exportables. The most relevant features are emphasized below.

Table 4.2—Variation in log of relative price of agricultural importables decomposed by source, 1980-87

Country	Period	Change in Domestic Price (1)	Change in Foreign Price (2)	Real Exchange Rate (3)	Direct Protection (4)
Malawi	1980-82	67.5	-16.5	22.1	61.8
	1982-85	-20.3	-5.8	18.3	-32.8
	1985-87	-10.3	-5.8	-5.0	0.5
	1980-87	36.9	-28.1	35.4	29.6
Tanzania	1980-82	17.4	-15.0	-20.0	52.5
	1982-85	7.6	79.4	-9.8	...
	1985-87	26.9	42.3	81.5	-96.9
	1980-87	51.9	-52.0	51.7	52.3
Zambia	1980-82	38.0	-43.6	7.2	74.4
	1982-85	-38.2	-15.0	64.7	-87.8
	1985-87	7.0	-27.5	39.0	-4.5
	1980-87	6.9	-86.1	110.9	-17.9
Zimbabwe	1980-82	0.9	-45.8	10.0	36.7
	1982-85	-13.3	-75.5	39.1	23.1
	1985-87	-7.5	2.1	-16.3	6.7
	1980-87	-19.9	-119.3	32.8	-66.5
All countries	1980-82	31.0	-30.2	4.8	56.3
	1982-85	-16.0	-43.9	28.1	-0.2
	1985-87	4.0	2.8	24.8	-23.5
	1980-87	19.0	-71.4	57.7	32.6

Notes: Numbers correspond to 100 times the change in the logs; therefore, figures can be approximately interpreted as percents. The domestic price is decomposed so that (1) = (2) + (3) + (4).

The "Leaning Against the Wind" Policy

The changes in the nominal protection rate (Table 4.1, column 4) tend to exhibit compensating movements vis-à-vis the movements in the international prices (column 2): when international prices go up, the nominal protection rate tends to go down and vice versa. This pattern is especially clear for Malawi and Tanzania. For Zambia and Zimbabwe, the pattern is weaker, the period 1982-85 being the main exception to the rule. In the case of Malawi the pattern is so strong that, as a result, the actual price incentives that agricultural exportables face move in opposite directions vis-à-vis the price signals coming from abroad. This strong countercyclical pattern in nominal protection rates is consistent with a pricing policy that sets domestic prices at levels that have little relation to border prices, so that no upward or downward movements in border prices are transmitted to domestic prices in any significant way. This countercyclical pattern in nominal protection rates for agricultural goods has been observed for a number of countries in a similar context (Schiff and Valdés 1992).

The implications of this implicit policy rule are clear and can be very important when considering this group of countries as a whole. Compare, for example, Malawi and Zimbabwe. In both countries, agricultural goods constitute a sizable share of total exports, so they face relatively similar foreign terms-of-trade shocks. This is reflected in a similar evolution of their real exchange rate. However, during the period under study, border prices of agricultural exportables declined more for Zimbabwe than for Malawi; indeed, the border price of exportables dropped by roughly 18 percentage points more for Zimbabwe. From an aggregate welfare point of view (taking the two countries as a unit), and assuming that at the beginning of the 1980s both countries faced a similar degree of intervention, incentives should have been more favorable for Malawi than for Zimbabwe because Malawi faced better export opportunities.[2] Yet, due to the stronger "leaning against the wind" policy followed by Malawi as compared with Zimbabwe, for the whole period (1980-87), price incentives in Zimbabwe for agricultural exportables improved by 10 percentage points more than in Malawi; exactly the opposite of what an open trade regime would have dictated.[3]

Negative Trend in Nominal Protection Rates

The countercyclical policy described above was not neutral in the long run, however. With the exception of Tanzania, the nominal protection rate tended to deteriorate during 1980-87. Although this does not necessarily imply the existence of a negative trend due to the selection biases of beginning and ending years, it is clear from the figures that such a trend exists for Zambia and Zimbabwe.

---

[2] Since Table 4.1 shows only the changes in the nominal protection, if in the starting year the protection rates had been especially high in Malawi, the conclusion might be false. However, this was not the case, as will be shown later in the chapter.

[3] Taking the two countries as a whole, optimally, resources should have moved in the direction of the less-depressed international prices.

It could be argued that this negative trend in nominal protection rates is the result of increasing transport costs due to the deterioration of infrastructure, lack of spare parts, and other related problems that these countries have been suffering in the last decade (see for example, World Bank 1989b, 5-7; World Bank 1990, 17; and Food Studies Group 1990, vi). Although this is sometimes the case, two additional considerations must be taken into account. First, even if the negative trend detected here could be explained partly by an increase in transport costs (which are assumed in this model to be constant in real terms), it is still an open question whether the increases in these transport costs are themselves a result of indirect government interventions in areas such as foreign-exchange allocation and public-sector involvement in the transport system. Second, no clear trend was detected for agricultural importables, so no potential alternative explanation based solely on increases in transport costs can be adopted.

Positive Trend in the Real Exchange Rate

For the four countries under study, the real exchange rate increased substantially during 1980-87 (Tables 4.1 and 4.2, column 3). It was mainly the upward movement in the real exchange rate that precluded a further drop in price incentives accruing from the negative tendency in the nominal protection rate.

As will be argued later, underlying the evolution of the real exchange rate, factors depending on world market conditions as well as factors related to government policies can be detected. Except for Tanzania, the important part of the increase in the real exchange rate took place during 1980-85, a period during which all four countries experienced negative shocks in their foreign terms of trade. It seems clear that different government policies in Tanzania, which resulted in high budget deficits and an undervalued nominal exchange rate, delayed an upward adjustment in the real exchange rate. Toward mid-1986, the Tanzanian government initiated a comprehensive structural-adjustment program that included a reduction in budget deficits, some liberalization in the foreign-trade sector and devaluations of the nominal exchange rate. The outcome of that program, in conjunction with the deterioration of the terms of trade that had already taken place in the first half of the 1980s, resulted in a substantial increase in the real exchange rate (a log increase of 0.82, or approximately 125 percent).

Table 4.2 shows the same decomposition for agricultural importables. The "leaning against the wind" policy is verified again, this time for Malawi, Tanzania, and Zimbabwe. Contrary to the case of exportables, however, no clear trend is observed in the rate of nominal protection. For the period 1980-87, the nominal protection rate increased for all countries except Zambia. This increase in the direct protection, in conjunction with the increase in the real exchange rate discussed above, produced an increase in the domestic price of importables in spite of a strong negative tendency in the corresponding international prices.

One of the conclusions to be drawn from Tables 4.1 and 4.2 is that, for each of the countries, a strong antitrade bias occurred. While the international relative prices of importables strongly deteriorated vis-à-vis the exportables, the contrary happened with the domestic relative prices. To take an example, for the case of Malawi the log of the foreign relative price of importables decreased by -0.28 for 1980-87 (Table 4.2, column 2), while the corresponding foreign price of exportables decreased by only 0.07, which implied an increase in the log of the foreign price of exportables relative to importables of 0.21 (approximately 23 percent). In spite of this, the corresponding domestic relative price decreased by approximately 0.66 (approximately -51 percent). Each of the four countries displayed a similar pattern for the whole period and for almost every subperiod that might be considered. One of the potential implications of this pattern of behavior is that productive resources may have moved in the wrong direction, that is, in the direction of more depressed international prices and away from better production opportunities. As a consequence, one may expect these policies to have implied a significant loss of foreign exchange.[4]

It is important to stress that the antitrade bias that intensified during this period probably had foreign-exchange losses associated with it. The importance of this stems from the fact that one of the underlying motives for discriminating in favor of importables, and against exportables, seems to be the need to achieve "food security," which is commonly identified with self-sufficiency in the production of basic food products. However, self-sufficiency is not the only way to achieve food security, and certainly not the optimal way. An alternative way to achieve food security is through trade, and this in turn requires export surpluses, which in these countries (perhaps with the exception of Zambia) are to be generated by the exportable activities in the

[4] Since Tables 4.1 and 4.2 show only the evolution of the protection rates but not the levels themselves, this assumes that at the beginning of the 1980s the protection levels of exportables and importables were either equal or already higher for the importables. The estimates of nominal protection rates presented later show that this was in fact the case.

agricultural sector. A possible reading then of the results of Tables 4.1 and 4.2 is that during the 1980s these countries pursued the objective of food security by putting too much emphasis on self-sufficiency, disregarding the alternative of trade, and hence intensifying an antitrade pattern that discriminates against production of exportables.[5] Although Tables 4.1 and 4.2 show only the evolution of the nominal protection rates and not the protection levels themselves, the antitrade pattern is confirmed in a later section that presents estimates of protection rates.

The results presented in Tables 4.1 and 4.2 shed some light also on the issue of the timing of economic reforms and the interaction with macro and micro reforms. One problem these countries face is coordination of the macroeconomic reforms leading to increases in the real exchange rate with the microeconomic reforms in the marketing of agricultural products. If inefficiencies in the marketing of agricultural products—involving parastatals and some cooperative unions—are not solved appropriately, there is the risk that the increases in the real exchange rate coming from macroeconomic reforms will not reach the farmers in the form of improved price incentives. This seems to have been the case with Tanzania during the first year and a half of the program of structural economic adjustment that started in mid-1986. Observe that in Tanzania, for the period 1985-87, the huge increase in the real exchange rate was completely offset by a decrease in direct protection to agriculture in both the exportable and importable sectors. This resulted in increases in domestic prices of modest size, compared with the increases in international prices and the real exchange rate. It seems clear then that macroeconomic reforms are a necessary but not a sufficient condition for better price incentives. Although lack of data precluded pursuit of this analysis for the years after 1987, the problem of imperfect transmission of real exchange rate increases to domestic agricultural prices still seems to be an important problem, at least in the Tanzanian case (World Bank 1989b, 8).

Another look at the evidence can be obtained by applying the variance operator to the log version of equation (4), assuming that the markup factor γ remains constant over the years. The countercyclical pattern is verified again by observing that for almost all the countries and cases, the

[5] As an argument in favor of self-sufficiency, one may think of some cases such as white maize where, given the particular condition of international market "thinness," a certain degree of protection may seem reasonable (although a first, best option would be a direct production subsidy). But this condition certainly does not apply to wheat or yellow corn.

covariance between direct intervention and border prices is negative, the exceptions being Zambia (exportables and importables) and agricultural exportables in Zimbabwe. A related consequence of this is that the variance of border prices appears substantially higher than the variance of domestic prices. The problem of real exchange rate increases not being transmitted to domestic agricultural prices appears clearly in Tanzania and Zambia, which exhibit a strong negative (normalized) covariance between the real exchange rate and the level of direct intervention. By the same token, the variability of the real exchange rate is between four and seven times higher than the variability of domestic prices. Finally, notice that in relation to agricultural exportables, Zimbabwe is a clear exception, since domestic prices exhibit a variability of similar size to border prices and the real exchange rate.

To obtain further insight into the structure of domestic policies affecting agriculture in these four countries, agricultural prices must be further decomposed to determine which part of the real exchange rate movements can be attributed to external shocks and which part is most probably the result of other macro policies pursued by these countries. Before doing this, however, a basic model for the determination of the real exchange rate is needed.

DETERMINANTS OF THE REAL EXCHANGE RATE, 1973-87

General Considerations

The main determinants of the real exchange rate are outlined in this section from a theoretical point of view, and their relevance for each country is highlighted. Next, the determinants of the real exchange rate are discussed with emphasis on the connection between the evolution of the real exchange rate and the evolution of the terms of trade. An econometric model, presented in the Appendix, was actually estimated for three of the countries and for the period 1973-87.

The real exchange rate is defined in the literature as the relative price between tradable and nontradable goods. In this model, as in most empirical research, this relative price is approximated by the ratio between foreign and national price levels. There has been a vast theoretical development attempting to explain the variations of the real exchange rate over time (Edwards 1989). To avoid reproducing all the theoretical arguments, the economic logic underlying the econometric model used in this study is briefly explained here.

The stochastic process followed by the real exchange rate is related to at least four other variables, namely, the foreign terms of trade of the economy, the current account, the trade policies (tariffs/subsidies to imports and exports), and the nominal exchange rate.

The Terms of Trade. A positive shock in the terms of trade is expected to be associated with a negative co-movement in the real exchange rate. The main effect of a positive shock in the terms of trade will be a positive wealth effect for the economy arising from better relative price of exports. The wealth effect will be more or less persistent, depending on the particular stochastic process followed by the terms of trade.[6] The wealth effect will induce a revaluation of all country-specific assets, thereby implying an upward movement in the price of nontradables, hence a decrease in the real exchange rate. It should be noted that this upward movement in the price of domestic assets does not have to wait for an increased domestic demand to take place; the movement can be fairly quick because the price of assets is a forward-looking variable. Mussa (1984) emphasizes this forward-looking nature of the real exchange rate.

The Current Account. In general, the theoretical literature has argued that no systematic pattern should be observed between the current account and the real exchange rate (Greenwood 1983). Yet the current account variable, or its counterpart, absorption over gross domestic product (GDP), has been used in many empirical studies of the real exchange rate in less-developed countries (Edwards 1989; Valdés 1990). Typically, the result is a positive association, that is, higher (lower) current account surpluses tend to be associated with higher (lower) levels of the real exchange rate. However, from a theoretical point of view, both the real exchange rate and the current account are considered endogenous variables; hence, the relation between the two may be volatile and depend crucially on the particular underlying process of the exogenous variables.

Roughly speaking, the crucial distinction relies on whether the current account is the result of supply or demand shocks. If the movement in the current account reflects mainly supply shocks affecting the whole economy (including the nontradable sector), the co-movement between the current account and the real exchange rate will tend to be positive: higher current-account surpluses will reflect positive domestic productivity shocks and,

---

[6] To take an extreme case, if the terms of trade follow a random walk, every change will be considered as permanent. On the other extreme, if the terms of trade follow a pure white noise process, all changes will be considered transitory.

hence, lower relative prices for nontradable goods. On the other hand, if the current account reflects mainly demand shocks (changes in government spending, for example), the relation will tend to be negative: a higher current-account deficit will be associated with an increased demand, which, in the presence of specific factors of production, will push the relative price of nontradables upward.

The above discussion suggests that although from a theoretical point of view no special relation prevails between the current account and the real exchange rate, the empirical findings for less-developed countries are consistent with a current-account stochastic process led mainly by demand shocks, most probably induced by an unstable government spending process.

Trade Policies. Important policy variables that affect the real exchange rate are trade restrictions and, in particular, the general level of import tariffs. With specific factors of production, and with substitution effects dominating potential income effects, it may be expected that an increase in import tariffs will be associated with a reduction in the real exchange rate (Sjaastad 1980). It is this negative effect in the real exchange rate, which is the main structural cause for the reduced export base (in terms of diversification), that is associated with countries following an inward-oriented growth strategy, characterized by high levels of protection of the industrial import-competing sectors. Unfortunately, no good proxy of this variable was available for these countries for the purposes of the econometric model presented in the Appendix. In the case of Tanzania, since mid-1986 an important effort has been made in terms of liberalizing foreign trade and the process for allocating foreign exchange. The lack of a good proxy for trade restrictions precluded the inclusion of Tanzania in the empirical implementation of a real exchange rate model (see Appendix).

The Nominal Exchange Rate. The question of whether shocks affecting the nominal exchange rate have any effects on the real rate is an old one, and it has reappeared in the more recent literature. More precisely, the question has often been posed in terms of the relation between the nominal exchange rate regime and the real exchange rate behavior. Mussa (1987) has pointed out that for the case of developed countries, the recent regime of a floating nominal exchange rate has been accompanied with a higher volatility in the real exchange rate. Mussa has argued that nominal rigidities may be responsible for this interaction between nominal and real variables, although the same result could be explained on different grounds. On the other hand, some authors have stressed the relation between the nominal and the real exchange rate when some degree of indexation prevails in the labor market (Aizenman and Frenkel 1985; Turnowsky 1987). Finally, Edwards (1989) has argued that, although in the long run the real exchange rate is not influenced by nominal variables, in the short run the level of the nominal rate may have real effects in the sense that it may speed up or delay the adjustment in the real rate that may be required in the presence of some real shock.

The relation between nominal and real variables is relevant for Tanzania, for example. Tanzania faced a pattern of deterioration of its foreign terms of trade similar to the other countries under study. However, the corresponding increase in the real exchange rate was delayed, in part due to a policy that insisted on the fixation of the nominal exchange rate during the early 1980s. More generally, these countries have experimented with a variety of exchange rate regimes; at times, the exchange rate has been pegged to the Special Drawing Rights (SDR), and more recently, to a basket of currencies reflecting the trade structure. Also, Zambia has experimented with a dual exchange rate system, one of the rates being determined in a semicontrolled auction market.

There is agreement in the literature that although changes in the nominal exchange rate may have an important influence on the behavior of the real exchange rate, in the long run the real exchange rate will be determined mostly by "fundamental" real variables only, such as the terms of trade, variables affecting the level of absorption of the economy (current account), and the general level of trade restrictions and distortions in general. The econometric model estimated in the Appendix is consistent with that view. It is important to stress this result, since policymakers too often try to target a high level of the real exchange rate without making the correspondent structural adjustments in the fundamentals of the economy. The long-run result of proceeding in that way is only a higher level of inflation, as successive devaluations raise the general price level without substantial long-run changes in relative prices. This remark is especially relevant for countries such as Tanzania, which has been relatively successful in inducing real exchange rate increases in the context of structural-adjustment programs. Future devaluations of the nominal exchange rate will be successful only to the extent that they continue to be associated with structural adjustments in the fundamentals (trade liberalization, continued reduction in budget deficits, and so forth).

The Real Exchange Rate in the Long Run

The evolution of the real exchange rate and the foreign terms of trade for the four countries for the period 1970-88 are shown in

Figures 4.1 to 4.4. It is apparent from the figures that the four countries under study have experienced similar long-run behavior in their terms of trade. For all countries, the average foreign terms of trade during the 1980s were between 30 percent and 60 percent lower than during the 1970s. For Malawi and Zimbabwe, the terms of trade show a long run decreasing trend, while for Tanzania and Zambia, a structural break can be identified in the mid-1970s and early 1980s, respectively. For Zambia, it is clear that the structural break corresponds to the oil crisis of the mid-1970s, which was associated with a strong drop in the price of copper that lasted more than a decade. In the cases of Malawi, Tanzania, and Zimbabwe, the effects of the oil crisis of the mid-1970s seem to have been partially neutralized by the

Figure 4.1—Real exchange rate and terms of trade in Malawi, 1970-88

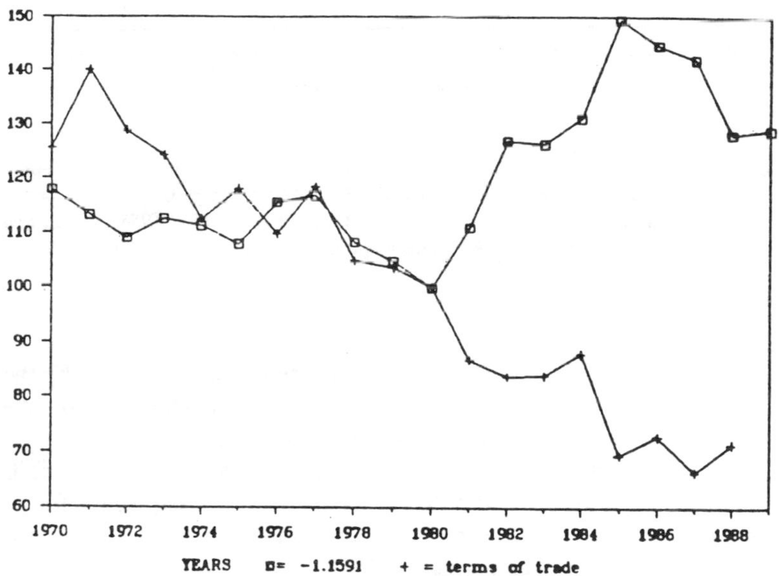

Sources: The real exchange rate is based on data from International Monetary Fund, *International Financial Statistics Yearbook* (Washington, D.C.: IMF, various years); terms of trade are from World Bank, *World Tables, 1988-1989 Edition* (Baltimore, Md., U.S.A.: Johns Hopkins University Press, 1989).

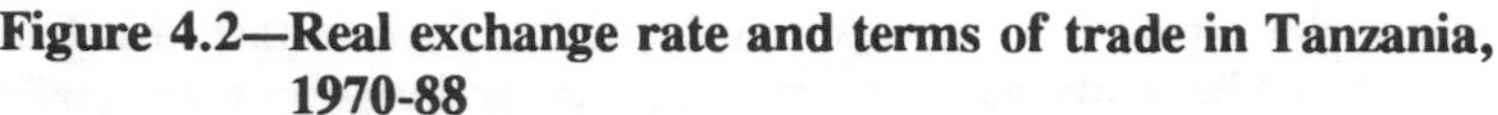

Figure 4.2—Real exchange rate and terms of trade in Tanzania, 1970-88

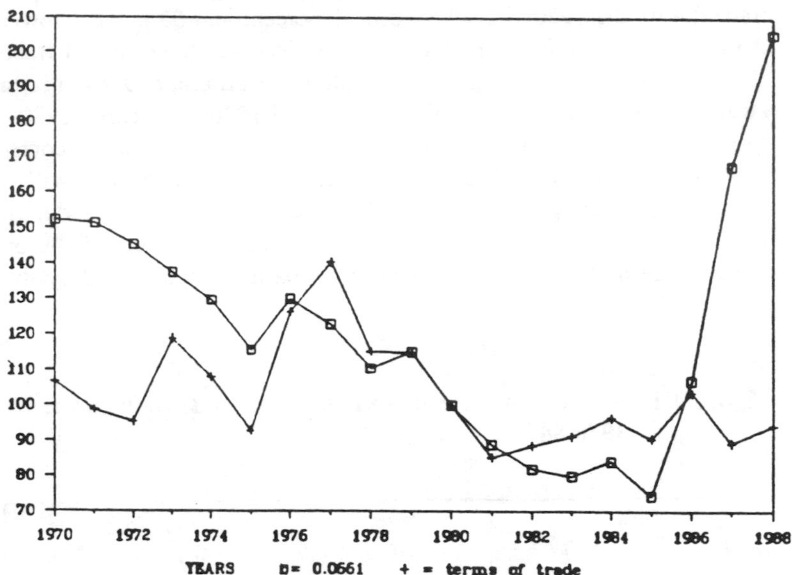

Sources: The real exchange rate is based on data from International Monetary Fund, *International Financial Statistics Yearbook* (Washington, D.C.: IMF, various years); terms of trade are from World Bank, *World Tables, 1988-1989 Edition* (Baltimore, Md., U.S.A.: Johns Hopkins University Press, 1989).

increase in the price of agricultural commodities that took place by the same date. However, during the 1980s, that neutralization did not take place, hence the continued decrease in the terms of trade.

As noted previously, in the long run the real exchange rate exhibits a clear negative relation with the terms of trade. Consequently, for all countries, the level of the real exchange rate during the 1980s was substantially higher than during the 1970s. In the cases of Malawi and Zimbabwe, the secular negative trend in the terms of trade practically mirrors an increasing trend in the real exchange rate—the relation is especially strong in the 1980s. At the bottom of all four figures, the slope coefficient of a simple least squares regression between the real exchange rate and the terms of trade is reported. The slope coefficient

Figure 4.3—Real exchange rate and terms of trade in Zambia, 1970-88

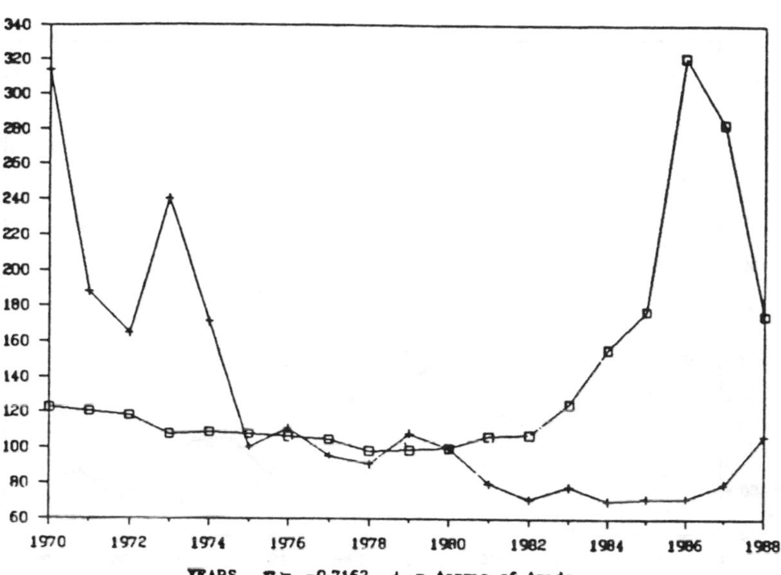

Sources: The real exchange rate is based on data from International Monetary Fund, *International Financial Statistics Yearbook* (Washington, D.C.: IMF, various years); terms of trade are from World Bank, *World Tables, 1988-1989 Edition* (Baltimore, Md., U.S.A.: Johns Hopkins University Press, 1989).

is negative for Malawi, Zambia, and Zimbabwe. For Tanzania, the coefficient is positive but close to zero (b = 0.06 in Figure 4.2).[7]

The last striking feature in Figures 4.1-4.4 concerns the possible timing of the adjustment of the real exchange rate to a deterioration of the terms of trade. Although for Malawi, Zambia, and Zimbabwe a substantial part of the deterioration in the terms of trade started taking place during the 1970s, the upward adjustment in the real exchange rate did not take place until the early 1980s. In the case of Tanzania, where

[7] The univariate regression between the real exchange rate and the terms of trade was made only for illustrative purposes. A more complete econometric model is presented in the Appendix. Note, however, that under the assumption that the terms of trade and the real exchange rate are cointegrated variables, the least squares regression is bound to pick up the long-run parameter associated with the terms of trade. Lack of degrees of freedom however, preclude pushing the argument of cointegration too far.

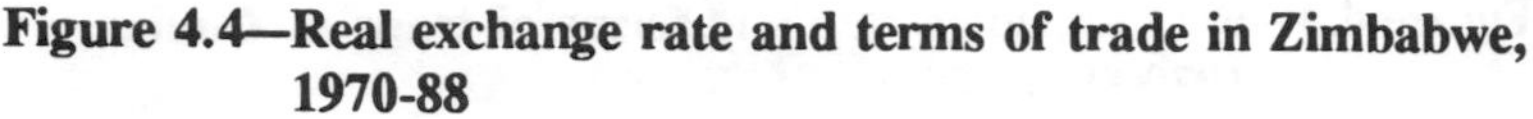

Figure 4.4—Real exchange rate and terms of trade in Zimbabwe, 1970-88

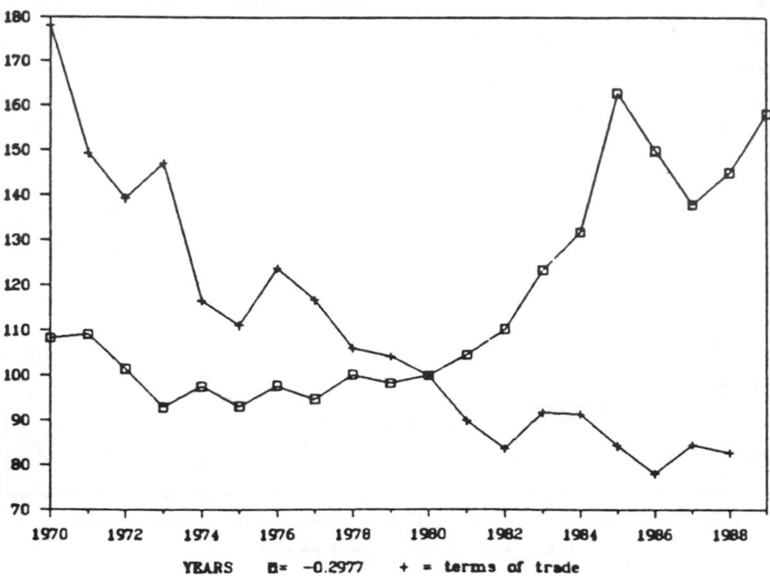

Sources: The real exchange rate is based on data from International Monetary Fund, *International Financial Statistics Yearbook* (Washington, D.C.: IMF, various years); terms of trade are from World Bank, *World Tables, 1988-1989 Edition* (Baltimore, Md., U.S.A.: Johns Hopkins University Press, 1989).

the substantial drop in the terms of trade took place in the early 1980s, the adjustment in the real exchange rate was delayed until the second half of the decade. The variables that explain the timing in the adjustment of the real exchange rate are related to the other determinants of the real exchange rate, which were outlined earlier. During the second half of the 1970s, the availability of international foreign credit as well as resources from donors allowed these countries to run important current-account deficits that allowed them to postpone upward adjustments in the real exchange rate. During the 1980s, credit conditions became tight, and a substantial effort to reduce current-account deficits was made. This effort was especially significant in Malawi and Zimbabwe. The reduction in the current-account deficits, in conjunction with the deterioration in the terms of trade that had already taken place, produced a significant upward adjustment in the real exchange rate.

In the case of Tanzania, the delay in the real exchange rate appears to be related to domestic policies pursuing an artificially low nominal exchange rate, and a highly distorted trade regime and exchange rate allocation system. The structural adjustments that started in the second half of the 1980s, and included a major realignment of the nominal exchange rate together with liberalization efforts in the foreign-trade sector, resulted in the increase in the real exchange rate that took place during 1986-88.

Empirical Results

The theoretical considerations and stylized facts presented above provide a framework that explains the behavior of the real exchange rate as the result of the joint evolution of exogenous factors to the economies, such as the terms of trade, and factors determined inside the economies, such as the level of expenditure and the nominal exchange rate. A pooled-data econometric model for Malawi, Zambia, and Zimbabwe that considers all these determinants is postulated in the Appendix. The main results are as follows:

- Current account surpluses (normalized by GDP) and terms of trade have a significant short-run impact on the real exchange rate, the point estimates of the short-run elasticities being 4.5 and -2.75 respectively.
- In the short run, nominal devaluations are fully transmitted to real exchange rate increases. The data appear better modeled, however, by assuming that nominal devaluations have no effect in the long run. This result is consistent with the view that in the long run, the real exchange rate is fully determined by fundamental variables only.
- The long-run effects of current-account surpluses and terms of trade on the real exchange rate—the elasticities being 1.59 and -0.76, respectively—are lower in magnitude than the short-run effects. Also, these elasticities are measured with less precision than the short-run ones. However, with a 90 percent confidence level, the sign of these estimates is certain.

In the following section these long-run elasticities are used to disentangle from the evolution of the real exchange rate which part can be attributed solely to external shocks like terms of trade changes and which part can probably be attributed to domestic policy variables.

AGRICULTURAL PRICES: A FURTHER DECOMPOSITION

An alternative decomposition of the changes in agricultural prices is presented in Table 4.3. The changes are decomposed according to whether they were originated by exogenous factors (terms of trade, border prices) or by endogenous or "policy-related" factors (nominal protection, current-account deficits, nominal exchange rate shocks). The details of the calculation are as follows.

Based on the model presented in the Appendix, the impact that the evolution of the foreign terms of trade had on the real exchange rate was computed. The implicit counterfactual scenario was one where the foreign terms of trade remained constant at their 1980 level. The predicted evolution of the real exchange rate due solely to the evolution of the terms of trade is presented in column 3 of Table 4.3. Adding this column to column 2 of Tables 4.1 and 4.2, one obtains the change in relative agricultural prices due to exogenous factors only, namely, changes in border prices and the induced effect on the real exchange rate due to the terms-of-trade shocks. These estimates are presented in columns 1 and 2 of Table 4.3. The changes in prices and real exchange rate that are not induced by exogenous factors were simply obtained as a residual between the first three columns of Table 4.3 and their actual counterparts in Tables 4.1 and 4.2.[8]

As an illustration, take the case of Zimbabwe during 1980-85. According to the actual changes in foreign terms of trade and international prices, the log of the relative price of agricultural importables would have declined by 0.885 (approximately –58 percent). However, specific and economywide government interventions counteracted that tendency by almost as much (0.761).

As shown in Table 4.3, the evolution of the foreign terms of trade implied a substantial increase in the real exchange rate. For the period 1980-87, the induced increase in the log of the real exchange rate ranged from 0.85 for Malawi to 0.32 for Tanzania, implying percentage changes of 133 percent and 37 percent, respectively. The explanation for this increase in the real exchange rate appears in Table 4.4: for all the countries under study, the terms of trade worsened drastically during 1980-87.

How did domestic policies react to these events? The last column of Table 4.3 shows that for Tanzania, Zambia, and Zimbabwe in the

---

[8] This approach carries a great deal of imprecision, since all the remaining stochastic error is attributed to the nonexogenous factors.

Table 4.3—Decomposition of price movements by external shocks and domestic policy components

	Log Price Changes Due to External Shocks			Log Price Changes Due to Domestic Policy		
Country/Period	Agri-cultural Imports (1)	Agri-cultural Exports (2)	Real Exchange Rate (3)	Agri-cultural Imports (4)	Agri-cultural Exports (5)	Real Exchange Rate (6)
Malawi						
1980-82	4.2	36.2	20.7	63.3	-52.0	1.5
1980-85	39.0	5.1	61.3	8.2	-14.9	-20.9
1980-87	57.2	78.1	85.3	-20.3	-107.7	-49.9
Tanzania						
1980-82	-0.8	-28.6	14.2	18.2	7.5	-34.3
1980-85	-69.2	-47.0	25.2	94.2	41.9	-55.0
1980-87	-20.3	-19.0	31.8	72.2	22.5	19.9
Zambia						
1980-82	-3.2	46.5	40.4	41.2	-39.1	-33.2
1980-85	9.4	2.7	68.1	-9.6	-64.3	3.8
1980-87	-7.3	76.8	78.8	14.2	-111.4	32.1
Zimbabwe						
1980-82	-24.7	6.7	21.1	25.2	5.4	-11.1
1980-85	-88.5	5.9	32.9	76.1	14.0	16.3
1980-87	-77.0	17.7	42.2	57.1	-37.7	-9.4
All countries						
1980-82	-6.1	15.2	24.1	37.1	-19.6	-19.3
1980-85	-27.3	-8.3	46.9	42.2	-5.8	-13.9
1980-87	-11.8	38.4	59.5	30.8	-58.6	-1.8

Notes: Figures correspond to 100 times the log changes and, therefore, can be approximately interpreted as percents. Columns (1) and (2) present the log of domestic price changes that would have occurred given full transmission of border prices to domestic prices in the same year. Column (3) represents the effect of the real exchange rate on changes in foreign terms of trade. Columns (4) through (6) represent the difference between the actual price changes and those explained by changes in foreign terms of trade presented in Columns (1) through (3). This was attributed to domestic policy.

early 1980s, the unfavorable evolution of the terms of trade was to some extent neutralized through other policies. In the case of Tanzania, the policies more than compensated for the deterioration of the terms of trade, and the real exchange rate actually decreased. For this country

Table 4.4—Macroeconomic variables and exchange rate distortion

Country	Period	Real Exchange Rate	Current Account over GDP[a]	Terms of Trade	Exchange Rate Distortion[b] (1)	(2)
Malawi	1980	100.0	-21.4	100.0	-57.1	...
	1980-82	112.6	-14.7	90.2	-46.8	-28.9
	1983-85	135.6	-5.7	80.4	-32.4	-19.9
	1986-87	145.3	-2.0	69.6	-26.6	-16.2
	1980-87	129.4	-8.2	81.4	-36.4	-22.4
Tanzania	1980	100.0	-8.4	100.0	-36.8	-22.6
	1980-82	90.2	-5.8	91.2	-32.6	-20.0
	1983-85	79.4	-3.0	92.7	-28.1	-17.2
	1986-87	116.3	-4.5	94.5	-30.6	-18.7
	1980-87	92.7	-4.4	92.6	-30.4	-18.6
Zambia	1980	100.0	-13.8	100.0	-45.4	-28.0
	1980-82	104.7	-16.0	83.6	-48.8	-30.2
	1983-85	152.8	-9.2	73.2	-38.0	-23.4
	1986-87	260.7	-11.4	74.2	-41.6	-25.6
	1980-87	161.7	-12.3	77.4	-43.0	-26.5
Zimbabwe	1980	100.0	-4.6	100.0	-30.6	-18.8
	1980-82	104.9	-8.3	91.2	-36.6	-22.5
	1983-85	139.4	-3.7	89.1	-29.3	-17.9
	1986-87	150.3	-0.2	82.3	-23.8	-14.4
	1980-87	129.2	-4.6	88.2	-30.7	-18.8

Sources: Terms of trade are from World Bank, *World Tables, 1988-1989 Edition* (Baltimore, Md., U.S.A.: Johns Hopkins University Press, 1989); current account figures are from International Monetary Fund, *International Financial Statistics Yearbook* (Washington, D.C.: IMF, various years).

[a] A negative figure corresponds to a deficit in the current account.

[b] The computation for column (1) uses a point estimate of 1.59 for the long-run elasticity of the real exchange rate with respect to the current account and an estimate of 0.78 for the elasticity of the real exchange rate with respect to the import tariff. The computation for column (2) uses a point estimate of 1.0 for the long-run elasticity of the real exchange rate with respect to the current account and an estimate of 0.7 for the elasticity of the real exchange rate with respect to the import tariff.

the countermovement in the real exchange rate lasted until 1985 and was probably a consequence of systematic misalignment in the nominal exchange rate. Later in the 1980s, Tanzania, Zambia, and Zimbabwe reversed their policies and validated important adjustments in the real exchange rate. In the case of Tanzania, with the structural-adjustment program of mid-1986, the situation changed, and the impact of terms of trade on the real exchange rate was validated through adjustments in the nominal exchange rate. Also, an additional effect must have operated through the impact of some trade liberalization and a more rational scheme for allocating foreign exchange.

For Malawi, the pattern was the opposite: at first it did not counteract the change induced by the terms of trade, but later some degree of neutralization took place.

For Malawi, Zambia, and Zimbabwe the results shown in Table 4.3 strongly suggest a policy bias against agricultural exportables. In these three cases, the exogenous factors would have resulted in substantial increases in the relative price of agricultural exportables; yet the outcome was a decrease in those prices, partly because of the neutralizing policies and partly because of the evolution of the nominal protection rate already discussed. On the other hand, Tanzania, Zambia, and Zimbabwe reveal a bias against importables: domestic policies induced upward movements in these prices at a time when world market conditions dictated the opposite. Finally, Malawi shows a case of general discrimination against agricultural tradables as a whole. Domestic prices of both agricultural exportables and importables decreased when world market conditions would have caused them to increase.

Three additional observations in regard to the results obtained here seem important. First, the determinants of the real exchange rate, which are different from the terms of trade, are labeled "domestic policy" factors in Table 4.3. A caveat is needed here. Clearly, the current-account variable is a result of policy decisions, but also of decisions by private agents. For one thing, if the decline in the terms of trade of the early 1980s had been seen as a transitory phenomenon, it would have been perfectly rational to run an increased current-account deficit as a way of smoothing consumption over time. From that perspective, the countermovement in the real exchange rate observed in some countries should not be considered a policy response. On the other hand, in these countries the capital market is quite repressed and underdeveloped, and numerous restrictions apply to capital transactions abroad by domestic private agents. So the government is the most important agent capable of running an excess of expenditure against a negative position in foreign assets. It is in this sense that the current account has been labeled within the domestic policy factors.

The second observation concerns the compensatory movements in the real exchange rate in the face of foreign terms-of-trade deterioration. Here the focus is on agricultural tradables. However, corresponding to those effects, opposite effects may be expected in the agricultural nontradables and probably even in the real wage if the nontradable sector is thought of as the most labor-intensive sector. To some extent, political considerations about the impact on real wages may underlie the compensatory response in the real exchange rate under adverse international circumstances.

Possibilities for Coordination. Concerning the issue of potential economic coordination between these countries, it must be said that the driving exogenous forces show a remarkably similar pattern in the four countries (see Figures 4.1-4.4 and Table 4.4). This common pattern reduces the possibility of coping with the real exchange rate risk (volatility) by means of, say, a regional compensatory foreign-exchange fund. However, it must be said also that Zambia shows a different pattern in her foreign terms of trade when longer periods of time are considered. This is due to the different composition of exports in Zambia, where copper dominates, and in the other three countries, where cash crops account for the largest share of exports. The asset-like properties of copper made the terms of trade in Zambia more volatile than in the other countries and produced a somewhat different evolution in the terms of trade. Indeed, toward 1988, Zambia was the only country that had higher terms of trade than in 1980, while in the early 1980s it was the country with the strongest deterioration in the terms of trade. So if there is some way of smoothing the impact of these exogenous shocks on the real exchange rate and hence on domestic relative prices without incurring major welfare losses, only an arrangement with Zambia on one side and the other three countries on the other side could have some chance of success.

EXCHANGE-RATE DISTORTION FROM DIRECT AND TOTAL INTERVENTIONS

The previous sections have decomposed the evolution of agricultural prices into different components. Emphasis has been made on the distinction between exogenous world-given factors as opposed to policy-related factors. Although such analysis yields significant insight into the structure of domestic policies and their implicit biases, it does not answer the related question of what adjustment in prices will take place if these countries pursue significant elimination of some basic trade and macro distortions in their economies.

To deal with this last question, a counterfactual scenario for macroeconomic and sector-specific policies is required. A useful benchmark for this analysis consists of the scenario where the nominal (direct) protection is removed simultaneously with the current-account imbalances and other trade-distorting policies. Specifically, the direct price intervention on agricultural product i is defined as

$$DIR_i = \frac{P_i}{\bar{P}_i^* E}, \tag{6}$$

where $\bar{P}_i^*$ is defined as the border price of good i once the transportation costs and other factors have been removed to make it comparable with the producer price. Let RER^* be the real exchange rate consistent with current-account balance and the removal of general trade distortions (import tariffs and export taxes in general), and define the real exchange-rate distortion RED analogously as

$$RED = \frac{RER}{RER^*} - 1. \tag{7}$$

Following closely the methodology proposed by Krueger, Schiff, and Valdés (1988) and recalling equation (1), the total price intervention, TOT, is defined similarly to the direct price intervention but in terms of relative prices with respect to the CPI, that is,

$$TOT_i = \frac{P_i}{P_i^*} - 1, \tag{8}$$

where P_i^* denotes the relative agricultural price that would prevail should the direct intervention and the real exchange-rate distortion be removed. With the definition of real exchange rate used here it is easy to verify that

$$1 + TOT_i = (1 + DIR_i)(1 + RED). \tag{9}$$

To compute the total and direct intervention, then, two basic calculations need to be done. First, the real exchange rate that would prevail under current-account balance and after the removal of general trade distortions needs to be estimated (RER^*). Second, some adjustment must be made to make border and producer prices comparable so as to get an estimate of the direct price interventions. It must be said at once that these estimations are subject to a much higher degree of uncertainty than any of the previous results presented in the other sections. First, the computation of the exchange-rate distortion will make an implicit assumption concerning strong exogeneity of the explanatory variables in the econometric model presented in the Appendix. Therefore, this estimation is subject in principle to the Lucas critique. Second, as will be seen later, the comparison of border prices with producer prices makes use of marketing and transport costs, which unfortunately are not available for all the countries. The results then must be taken with care, and as indicative of very general tendencies only. The possible direction of the biases will be given precisely whenever possible.

The last two columns of Table 4.4 present the real exchange-rate distortion calculated for all the countries and for different subperiods during 1980-87. The first computation of the distortion uses the point estimate of 1.59 for the long-run elasticity of the real exchange rate with respect to the current account (Appendix, Table 4.8). Also, it uses an estimate of 0.78 for the elasticity of the real exchange rate with respect to import tariffs. This estimate was taken from Mlambo (1989) and corresponds to an estimate based on Zimbabwe's data. The second estimation of the real exchange-rate distortion is more "conservative." Due to the imprecision of the estimate of the long-run elasticity of the current account, the second computation used an elasticity of 1 instead of 1.59. The probability that the actual elasticity is greater than this number is around 0.68, hence the term "conservative."[9] Also, this second estimation of the distortion used a coefficient of 0.7 for the elasticity with respect to import tariffs. This last number is taken from an average of other studies quoted in Valdés (1990).

Both computations of the real exchange-rate distortion required an estimate of the average level of trade distortions in the economy, approximated by the concept of "uniform equivalent tariff" developed in Sjaastad (1980). Unfortunately, no such estimate was available. On the other hand, trying to estimate the general tariff level by using data from the revenues from import duties (IMF, various years a) compared with the level of total imports is far from appropriate due to the extensive use of quantitative restrictions, advance deposits, and other nontariff instruments by these countries (IMF, various years c). In fact, the computation of the tariff by this method yielded stable estimates for all the years in the range of 10-20 percent for all the countries except Tanzania. For Tanzania, the estimates were in the range of 5-10

---

[9] This probability is calculated from the standard deviations presented in the Appendix, Table 4.8.

percent. However, from the reports on trade restrictions (IMF, various years c), it appeared that the lower tariff estimates for Tanzania were probably the result of a more intensive use of quantitative restrictions by this country.

The recurrent use of different forms of export retention schemes by Tanzania, Zambia, and Zimbabwe suggests that the level of quantitative restrictions on imports is fairly significant. In this study, conservative values were used for the average tariff levels, so the level of real exchange-rate distortion will, in general, be biased downward. The first estimate used 30 percent for the equivalent tariff level, while the second (conservative) used 20 percent. These estimates appear in line (although somewhat lower) with those used by DeRosa (1992) in his study of Sub-Saharan Africa.

In general, the level of exchange-rate distortion has fluctuated between 15-40 percent, depending on the country, the year and the particular computation used (Table 4.4). In other words, the removal of import tariffs and a balanced current account would result in an increase in the real exchange rate of 15-40 percent. In general, the exchange-rate distortion has been fairly similar in Malawi, Tanzania, and Zimbabwe, while in Zambia it has consistently been around 5 to 10 percentage points higher. For the case of Zimbabwe, the estimates in this study appear in line with the exchange rate subvaluation quoted in the Food Studies Group (1990, i). Note also that the exchange-rate distortion in the four countries has shown a tendency to decline over the years due to significant adjustments of the current-account imbalances. These adjustments have been particularly important in the case of Malawi. Finally, it is important to note that the exchange-rate distortion was computed with long-run elasticity estimates. As a consequence, the dynamic issues have been overlooked. For example, in the case of Zambia, it is very possible that the distortion estimate could be wrong for the period 1985-87. The high level of the real exchange rate for this subperiod suggests an adjustment in the current account that has not been completed.

Turning now to the estimation of the direct price intervention, the adjustments in the border prices were taken basically from the estimations by Jansen (1982, 1988), who calculated them for Zambia and Zimbabwe, respectively. The same markup factors were used for the other countries. As a consequence, the results should be taken as a very rough approximation only. Unfortunately, the quota policy on some exports, the extended operations of parastatals in many cases, and the possible differences in transport costs for the different countries, make a precise estimation of the direct protection rate a very difficult task. Further, Tanzania's and Zambia's practice of panterritorial pricing policies for some products during some years, complicates the very concept of an aggregate protection rate. Hence, the results obtained here should be interpreted cautiously.

Table 4.5 shows the basic results. Three main stylized facts seem clear. First, there is a marked tendency to protect agricultural importables more than the exportables; the antitrade bias is confirmed again. A useful way to quantify this antitrade bias is to divide one plus the average direct protection rate to importables by the corresponding expression for exportables. The result gives the average percentage by which relative prices of agricultural exportables vis-à-vis agricultural importables would increase if the specific sector pricing policies were removed. Using the averages of Table 4.5, this exercise gives the following results for the antitrade bias of direct price interventions during 1980-87: Malawi, 1.86; Tanzania, 1.95; Zambia, 1.41; and Zimbabwe, 1.88.

These figures mean that the removal of direct price intervention would result in relative price increases of exportables relative to importables of 86 percent, 95 percent, 41 percent, and 88 percent for Malawi, Tanzania, Zambia and Zimbabwe, respectively. It is important to notice that although the levels of direct protection in each country may be subject to bias due to the imprecise estimation of reference marketing and transport costs, this bias should be smaller in the above estimation of the antitrade index because the biases in the numerator and denominator of the index would tend to cancel each other out.

The second feature that appears in Table 4.5 is that Tanzania, Zambia, and Zimbabwe show a marked difference from Malawi in the sense that they protect the agricultural sector more (or disprotect less), the rough difference being at least 30 percentage points. Again, although this result could be biased due to differences in transport and marketing costs across countries, the difference in direct protection appears to be strong enough to overcome those possible biases, at least at a qualitative level.

The third feature is that Zambia and Zimbabwe appear to be the countries with the lowest antitrade bias and the highest direct protection rates to agriculture, respectively.

It is important to point out that this last stylized fact fits closely with the political economy underlying price interventions in these countries. On one hand, for Zambia, agricultural exportables are not important as a share of total exports; copper accounts for as much as 90 percent of total exports (Jansen 1988). A general tendency of policymakers is to put the greater burden of the tax on the largest

Table 4.5—Direct protection rate estimates, selected years

Country	Period	Exportable Products			Importable Products	
		Tobacco	Tea[a]		Wheat	Maize
Malawi	1980-82	-73	n.a.		-43	-44
	1983-85	-62	n.a.		-48	-41
	1986-87	-71	n.a.		-30	-26
	1980-87	-68	n.a.		-42	-38
		Coffee	Cotton		Rice	Maize
Tanzania	1980-82	-38	-31		28	-14
	1983-85	-7	20		101	106
	1986-87	-49	-12		115	53
	1980-87	-29	-7		77	48
		Tobacco	Cotton		Wheat	Maize
Zambia	1980-82	10	0		112	9
	1983-85	23	-32		66	-2
	1986-87	n.a.	n.a.		-4	-33
	1980-87	0	-16		66	-5
		Tobacco	Cotton	Maize	Wheat	Rice
Zimbabwe	1980-82	18	-20	-14	54	23
	1983-85	-2	-46	13	80	88
	1986-87	-15	-37	1	57	122
	1980-87	2	-34	0	64	72

Notes: n.a. means not available. The direct price intervention on agricultural product i is defined as

$$DIR_i = \frac{P_i}{\overline{P}_i^* E},$$

where $\overline{P}_i^*$ is the border price of good i once the transportation costs and other factors have been removed to make it comparable with the producer price.

[a] No reliable adjustment to the border price could be made in the case of tea.

sector—typically, the one with the highest comparative advantage.[10] In that sense, the low level of implicit export taxes in Zambia is an expected result. On the other hand, in Zimbabwe the larger-scale commercial farm sector has traditionally been influential in the policy process (World Bank 1984) and would be expected to lobby for a high

[10] Schiff and Valdés (1992) present some stylized facts regarding price interventions that are in agreement with this pattern.

protection to agricultural goods.

Since in most of the cases considered here, the direct protection rate is an implicit outcome of the pricing policies of the many parastatals involved in the marketing of agricultural products, it is important to reconcile, at least qualitatively, the results on direct protection obtained here with other pieces of evidence concerning the operation of parastatals. In most of these countries, the parastatals run high deficits which require a continuous stream of subsidies from the central government. It might seem that those deficits are in contradiction with some of the results obtained here. For example, in Tanzania and Zimbabwe, the present study obtains negative protection to exportables and positive protection to importables for almost all products and years. These direct protection rates could be seen as implicit export and import taxes; hence, they should generate revenue and not cause major deficits. In this respect, two observations need to be made.

First, the protection rates considered here correspond to protection given to producers and not to consumers. The parastatals protect producers of importables for food-security reasons and set a low price to consumers, the net result being a deficit. That seems to be the case with wheat in Zimbabwe, for example. Second, it must be emphasized again that the measures of protection presented here were constructed based on reference (not actual) transportation and marketing costs. There seems to be agreement that a number of inefficiencies are present in the marketing process conducted by the parastatals and also in their dealings with cooperative unions, for example. Those inefficiencies have resulted in marketing costs far above cost levels arising from more competitive and efficient marketing processes. In that sense, the potential revenue that could be generated by direct intervention policies is not realized due to an inefficient marketing process.

The total protection rate, calculated from the previous two tables by using equation (9) above, is shown in Table 4.6. The exchange-rate distortion used in this computation corresponds to the first estimate in Table 4.4. On average, Table 4.6 shows that Zimbabwe is the country with the highest total protection rate toward agriculture (the lowest disprotection), while Malawi is the opposite. The most striking result shown in this table is the strong total disprotection suffered by agricultural exportables. Taking an average over all products and years, the total disprotection rate ranges from -28 percent to Zimbabwe to -80 percent to Malawi. The average for importables and exportables together yields a total protection of -10 percent for Zimbabwe and -68 percent for Malawi. Thus, the removal of general trade distortions,

Table 4.6—Total protection rate estimates, selected years

Country	Period	Exportable Products			Importable Products	
		Tobacco	Tea[a]		Wheat	Maize
Malawi	1980-82	-86	n.a.		-70	-70
	1983-85	-74	n.a.		-65	-60
	1986-87	-79	n.a.		-49	-46
	1980-87	-80	n.a.		-63	-61
		Coffee	Cotton		Rice	Maize
Tanzania	1980-82	-58	-54		-14	-42
	1983-85	-33	-14		45	48
	1986-87	-65	-39		49	6
	1980-87	-51	-35		23	3
		Tobacco	Cotton		Wheat	Maize
Zambia	1980-82	-44	-49		9	-44
	1983-85	-24	-58		3	-39
	1986-87	-71	n.a.		-44	-61
	1980-87	-43	-52		-6	-46
		Tobacco	Cotton	Maize	Wheat	Rice
Zimbabwe	1980-82	-25	-49	-14	-3	-22
	1983-85	-30	-62	13	27	33
	1986-87	-35	-52	1	19	69
	1980-87	-29	-54	0	14	19

Notes: n.a. means not available. The total price intervention, *TOT*, is defined similarly to the direct price intervention but in terms of relative prices with respect to the *CPI*, that is,

$$TOT_i = \frac{P_i}{P_i^*} - 1 ,$$

where P_i^* denotes the relative agricultural price that would prevail should the direct intervention and the real exchange distortion be removed.

[a] No reliable adjustment to the border price could be made in the case of tea.

macro distortions, and sector-specific policies would imply an increase in relative agricultural prices of the order of 10 percent to 68 percent, depending on the country.

The difference in total protection between Zimbabwe and Malawi is strong enough to overcome possible biases due to errors of measurement in the different steps of the estimation. In the middle of the spectrum are Zambia and Tanzania with average total protection rates of −37 percent and −15 percent, respectively. Note also that in many cases the direct protection given to importables was sufficiently strong that total protection remained positive after the exchange-rate distortion was adjusted for. Finally, observe that if the conservative computation of the exchange-rate distortion had been used instead, the above conclusions would have remained roughly unchanged. In particular, it would still be the case that, on average, all the countries except Zimbabwe would exhibit a negative total protection rate. For the particular case of Zimbabwe, the average for all the years and goods would be around zero.

To conclude this section it is important, first, to assess the relation between the results obtained in this study and the actual policies followed by these countries and, second, to determine how these results can enlighten future policy prescriptions.

The decrease in the exchange-rate distortion in Malawi and Zimbabwe during 1980-87 seems consistent with the main macroeconomic developments in these countries. In the particular case of Malawi, the decrease in the distortion appears in line with the efforts made under the structural-adjustment program for the period 1982-86 with assistance from the International Development Association and the International Monetary Fund. The reductions in the budget deficits allowed the economy to reduce the high level of current account deficits of the early 1980s and brought an increase in the real exchange rate and a consequent reduction in the level of the distortion measured here. Note that the consequence of these adjustments for the case of importable products was that toward 1986-87 the total disprotection rate was 20-25 points lower than in the early 1980s. However, in spite of this progress, the total protection rate still appears very negative, for importables and exportables alike, in the last years of the sample.

The case of Tanzania has been already discussed. Note, however, that the estimate of the real exchange-rate distortion does not reveal a significant decline for the period 1986-87. This is probably a weakness of the methodology, since the computation of the real exchange-rate distortion did not consider the foreign trade liberalization measures adopted since mid-1986. Taking this into account, and assuming that the partial trade liberalization amounted to a reduction of the order of 10-15 percent in the global level of protection, the level of exchange-rate distortion for these years would be in the range of 5-10 percentage points less. However, as discussed before, Tanzania also has to solve the problems of inefficiencies and distortions in the marketing process of agricultural products that have precluded the increases in the real exchange rate from being transmitted to farmer price incentives in a more substantial way. A similar observation applies to Zambia.

SUMMARY AND CONCLUSIONS

The evolution of agricultural prices during 1980-87 has been decomposed into exogenous and domestic policy components, and an estimation of the real exchange-rate distortion and the total protection toward agriculture has been presented for the four countries. The main conclusions are as follows.

Domestic policies toward agriculture during 1980-87 had a strong antitrade bias: exportable products tended to be disprotected and importable products tended to be protected. The removal of sector-specific policies would have implied that the relative price of agricultural exportables vis-à-vis agricultural importables would have increased by a percentage between 41 percent and 95 percent depending on the country.

The total disprotection rate for agricultural exportables was very high and varied, on average, from −28 percent for Zimbabwe to −80 percent for Malawi.

Over time, domestic policies showed a countercyclical component, especially in regard to exportables. High border prices tend to be offset by increases in the disprotection rate and vice versa.

During 1980-87, the real exchange rate increased. This increase averted a drastic deterioration in the domestic price of exportables. However, for some countries (mainly Tanzania and Zambia), the real exchange rate increases are far from being transmitted in the form of substantially higher prices to the farmers.

The increase in the real exchange rate seems to be strongly associated with a secular deterioration in the terms of trade and with positive adjustments in the current-account imbalances. Also, the policies regarding nominal devaluations in the exchange rate seemed to have played an important short-run role in the determination of the real exchange rate.

The level of the real exchange-rate distortion ranges between −15 percent and −40 percent, and it is fairly similar across the countries.

The antitrade bias seems to be lower in Zambia. Also, Malawi shows the strongest disprotection rate against agriculture.

Based on these conclusions, several suggestions can be made. First, the antitrade bias in conjunction with the countercyclical policies implies that resources have flowed in the wrong direction, that is, in the direction of more depressed international prices. This implies an unknown loss of foreign currency for these countries. In terms of the countercyclical policies at least, it would be desirable to look for more efficient means of price stabilization.

Second, if the antitrade bias is partly a consequence of a self-sufficiency objective, further economic integration between these countries may contribute to diminishing that bias and hence to a net foreign-currency saving. On the other hand, the removal of the antitrade bias seems a necessary precondition for successful economic integration.

Third, the similar evolution in the terms of trade suggests that the possibilities for neutralizing real exchange rate risk among these countries are not very high. Arrangements with Zambia on one side and the other three countries on the other may have a chance of success.

Finally, looking beyond the issue of economic integration among southern African countries, the total disprotection toward agricultural exportables is sufficiently high to make a consideration of price adjustment in the direction of international incentives worth studying. The removal of trade and macro distortions (with the consequent increase in the real exchange rate) and the reduction in the taxation of agricultural exportables seem necessary conditions for the future expansion of agricultural exports in these four countries. This may involve a reform in the agricultural marketing sector so that increases in the real exchange rate are effectively transmitted into better price incentives.

APPENDIX: AN ECONOMETRIC MODEL FOR THE REAL EXCHANGE RATE

The discussion of determinants of the real exchange rate (*RER*) in the 1973-87 period suggests the following statistical model:

$$\overline{RER}_{it} = \alpha_0 + \sum_{j=0}^{j=N} (\alpha_{1t-j}\,\overline{E}_{it-j} + \alpha_{2t-j} ca_{it-j} + \alpha_{3t-j}\,\overline{tt}_{it-j}) + \beta \overline{RER}_{it-1} + \epsilon_{it}\,, \qquad (10)$$

where $\overline{RER}_{it}$ = log of the real exchange rate for country i at time t, defined as the ratio of international and domestic prices;

E_{it} = corresponding log of the nominal exchange rate;

ca = current-account surplus (deficit = $-ca$) as a proportion of total GDP;

$\overline{tt}_{it}$ = log of the foreign terms of trade expressed as price of exports over price of imports; and

ϵ = random error.

If equation (10) represents a reasonable, though parsimonious, approximation for the true process of the *RER*, it can be expected to yield nonexplosive dynamics so that the absolute value of β will be between zero and one. For any explanatory variable (k = 1...3), the long-run impact of that variable on the $\overline{RER}$ will be given by

$$\frac{\sum_{j=0}^{j=N} \alpha_{kt-j}}{1 - \beta}. \qquad (11)$$

According to the discussion on the determinants of the *RER*,

$$\sum_{j=0}^{j=N} \alpha_{2t-j} > 0 \quad \sum_{j=0}^{j=N} \alpha_{3t-j} < 0. \qquad (12)$$

If the process for the nominal exchange rate affects the real rate the coefficients α_{lt-j} are expected to be different from zero. If, in the long run, nominal variables do not affect real ones, the corresponding sum of coefficients should be zero. Finally, note that equation (10) can usually be reparameterized to yield an error-correction representation for the *RER* (Hendry and Richard 1982). This will not be pursued here.

RESULTS OF ESTIMATION

Equation (10) was estimated by minimum least squares, pooling cross-sectional data for Malawi, Zambia, and Zimbabwe, and time series data for the period 1973-87. Annual data were used and a lag of one year proved to be enough for capturing the essential dynamics of the real exchange rate (that is, N = 2 in equation [10]). Dummies were included to test for systematic differences in levels across countries, but they were not significant. Also, systematic differences in error variances across countries did not appear. Finally, the inclusion of Tanzania deteriorated the global properties of the model, suggesting that for this country, either all coefficients were different or additional variables needed to be included. In either case, the lack of degrees of freedom and the lack of additional information, or both, made an independent estimation infeasible. For the selected set of countries (Tanzania excluded), the results of the estimation appear in Table 4.7.

Note first that the nominal exchange rate was included as a first difference in the set of explanatory variables. This amounts to saying that nominal exchange rate shocks affect the real rate only in the short run (within the year), while in the long run the real exchange rate is affected only by the terms of trade and the evolution of the current account (that is, for the nominal rate, the coefficients in equation [11] total zero). From a theoretical point of view, this restriction implies that the impact of the nominal rate on the real rate might be due to short-term nominal rigidities that disappear in the long run. When this restriction was not imposed, the properties of the error term in the estimated equation strongly deteriorated (D.W. below 1.5) and the absolute value of the lagged endogenous variable was far above one

Table 4.7—Real exchange rate equation, 1973-87

Variable	Point Estimate	t-Statistic
$\Delta\bar{E}$ 1.92	3.09	
ca_t	4.49	2.84
ca_{t-1}	-2.24	-1.39
$\bar{\pi}_t$ -2.75	-2.91	
$\bar{\pi}_{t-1}$ 1.68	2.01	
$\overline{RER}_{t-1}$	-0.41	-0.81
trend	-0.07	-1.52

$\bar{R}^2 = 0.36$
$D.W. = 1.83$

implying an odd dynamic pattern (explosive). The model imposing the above restriction then seems far more plausible and appears in agreement with most of the literature.

The point estimate of the nominal exchange rate shock suggests the possibility of an "overshooting," that is, a nominal devaluation of 10 percent having a short-run impact of roughly 20 percent on the real rate. A possible rationalization of this result may be related with the potentially contractionary effects of a devaluation, an issue that has been addressed in the literature.

Note that the short-run effects of the current account and terms of trade shocks are both significant and have the expected signs, with point elasticity estimates of 4.49 and –2.75 respectively. With respect to the long-run effects, they also have the expected sign (the sum of coefficients of the current and lagged explanatory variables) but, due to the high variance of the estimate of the coefficient accompanying the lagged endogenous variable, the long-run elasticities are less precisely estimated. In fact, using the matrix of variances and covariances of the estimated coefficients, the asymptotic distribution of the long-run elasticities defined by equation (11) can be calculated. The results are presented in Table 4.8.[11]

Table 4.8—Real exchange rate estimation using long-run elasticities

Variable	Point Estimate	Standard Error	t-test	Probability That the Parameter is Greater Than Zero
Terms of trade	–0.76	0.56	1.35	0.08
Current account	1.59	1.22	1.30	0.90

[11] To compute the t tests of Table 4.8, the distribution has to be obtained for $\tilde{\phi} = \tilde{\alpha}/1 - \tilde{\beta}$, where $\tilde{\alpha}$ corresponds to the sum of current and lagged estimated coefficients. The distribution of $\tilde{\alpha}$ can be trivially computed based on the matrix of variances and covariances of the estimated coefficients. For the asymptotic distribution of ϕ, the Taylor expansion is considered:

$$T^{1/2}(\tilde{\phi} - \phi) = T^{1/2}\frac{\partial\phi}{\partial\alpha}(\tilde{\alpha} - \alpha) + T^{1/2}\frac{\partial\phi}{\partial\beta}(\tilde{\beta} - \beta),$$

and the remainder of the above expression can be shown to converge asymptotically to zero, under standard regularity conditions. Since the two terms in the above expression follow known normal distributions with known covariances, the distribution of $\tilde{\phi}$ is readily obtainable.

From the estimates of Table 4.8 it is observed that an exogenous decline of 20 percent in the terms of trade, roughly equivalent to the average of the actual decline observed during the period under study (see Table 4.4), should imply a long-run upward movement in the real exchange rate of approximately 15 percent. On the other hand, and again using the point estimates, to correct a current-account deficit amounting to, say, 10 percent of GDP, the real exchange rate would have to increase by approximately 16 percent.

The above calculations give a rough idea of the basic magnitudes involved. However, these estimates must be interpreted with caution. The asymptotic standard errors of the long-run estimates are high, yielding in turn a low t-value. It must be said, however, that although the t-value is low, the signs of the respective parameters of interest are those that would be expected. For example, taking a Bayesian standpoint and starting with a uniform prior ranging over all the real line for all the parameters under study (improper prior reflecting complete ignorance), the posterior distribution of the parameters will match the distribution of the maximum likelihood estimates that appear in Table 4.8. Therefore, it is possible to compute the probability that the parameter is greater than zero for both cases. The last column of Table 4.8 presents such a calculation. The probability that a positive shock in the terms of trade increases the real exchange rate is reasonably low (0.08), while the probability that an increase in the current-account surplus will increase the real exchange rate is fairly high (0.9). Although the point estimates of the long-run elasticities tend to be imprecise, the terms of trade and the current account are most probably associated with the real exchange rate in the expected way.

To conclude the analysis of the econometric results, observe that the corrected R^2 is low, suggesting that a great deal of variation is left unexplained by the model. On the other hand, the Durbin Watson statistic allows acceptance of the null hypothesis of no autocorrelation of residuals. This test, however, is biased toward acceptance of the null due to the presence of the lagged endogenous variable in the set of explanatory variables. On the other hand, the alternative h-Durbin test cannot be computed in this case.[12]

[12] The number of observations times the variance of the estimate of the coefficient corresponding to the lagged endogenous variable was bigger than one.

REFERENCES

Aizenman, J., and J. A. Frenkel. 1985. Optimal wage indexation, foreign exchange intervention and monetary policy. *American Economic Review* 75 (3): 402-423.

Barnett, W. 1982. *Divisia indices*. Vol. 2, *Encyclopedia of statistical sciences*, ed. S. Kotz and N. Johnson, New York: Wiley.

DeRosa, D. A. 1992. Protection and export performance in Sub-Saharan Africa. *Weltwirtschaftliches Archiv* 128 (1): 88-124.

Edwards, S. 1989. *Real exchange rates, devaluation and adjustment in less developed countries*. Cambridge, Mass., U.S.A.: MIT Press.

Food Studies Group. 1990. Agricultural marketing and pricing in Zimbabwe. Background paper to the World Bank's Agricultural Sector Memorandum. Queen Elizabeth House, University of Oxford, Oxford, U.K.

Greenwood, J. 1983. Expectations, exchange rates and the current account. *Journal of Monetary Economics* 12 (4): 543-569.

Hendry, D., and J. F. Richard. 1982. Formulation of empirical models in dynamic econometrics. *Journal of Econometrics* 20 (1): 3-33.

IMF (International Monetary Fund). 1989. *Direction of trade statistics yearbook*. Washington, D.C.: IMF.

_____. Various years a. *Government financial statistics yearbook*. Washington, D.C.: IMF.

_____. Various years b. *International financial statistics yearbook*. Washington, D.C.: IMF.

_____. Various years c. *Report on exchange arrangements and trade restrictions yearbook*. Washington, D.C.: IMF.

Jansen, D. 1982. Agricultural prices and subsidies in Zimbabwe: Benefits, costs and trade-offs. International Food Policy Research Institute, Washington, D.C. Mimeo.

_____. 1988. *Trade, exchange rate, and agricultural pricing policies in Zambia*. A World Bank Comparative Study in The Political Economy of Agricultural Pricing Policy series. Washington D.C.: World Bank.

Krueger, A. O., M. Schiff, and A. Valdés. 1988. Agricultural incentives in developing countries: Measuring the effect of sectoral and economywide policies. *World Bank Economic Review* 2 (3): 255-271.

Masters, W. A. 1991. Trade policy and agriculture: Measuring the real exchange rate in Zimbabwe. *Quarterly Journal of International Agriculture* 30 (1): 21-36.

Mlambo, K. 1989. Exchange rate overvaluation and agricultural performance in Zimbabwe. In *Household and national food security in southern Africa*, ed. M. Godfrey and R. Bernstein, 243-258. Harare: University of Zimbabwe/Michigan State University.

Mussa, M. 1984. The theory of exchange rate determination. In *Exchange rate theory and practice*, ed. J. Bilson and R. C. Marston, 13-78. Chicago: University of Chicago Press.

_____. 1987. Nominal exchange rate regimes and the behavior of real exchange rates: Evidence and implications. In *Real business cycles, real exchange rates and actual policies*, ed. K. Brunner and A. Meltzer. Carnegie Rochester Conference Series on Public Policy. New York: North Holland.

Schiff, M., and A. Valdés. 1992. *A synthesis of the economies in developing countries*. Vol. 4, *The political economy of agricultural pricing policy*. Baltimore, Md., U.S.A.: Johns Hopkins University Press.

Sjaastad, L. 1980. Commercial policy, true tariffs, and relative prices. In *Current issues in commercial policy and diplomacy*, ed. J. Black and B. Hindley. New York: St. Martin's Press.

Turnowsky, S. 1987. Optimal monetary policy and wage indexation under alternative disturbances and information structures. *Journal of Money, Credit and Banking* 19 (2): 157-180.

Valdés, A. 1990. Preliminary estimates of exchange rate misalignment for Malawi, Zambia and Zimbabwe. Paper prepared for the IFPRI/SADCC Policy Workshop "Trade in Agricultural Products among SADCC Countries," 27-28 February, Harare, Zimbabwe.

World Bank. 1984. *Zimbabwe, agricultural marketing and input supply*. Working Paper. Subsector Memorandum, Eastern Africa Projects, Southern Agriculture Division. Washington, D.C.: World Bank.

_____. 1989a. Report to the consultative group for Tanzania on the government's economic and social policy reform program. Southern Africa Department, World Bank, Washington, D.C. Mimeo.

_____. 1989b. *World tables, 1988-1989 Edition*. Baltimore, Md., U.S.A.: Johns Hopkins University Press.

_____. 1990. Report and recommendation of the president of the IDA to the executive directors on a proposed credit to the Republic of Malawi. World Bank, Washington, D.C. Mimeo.

Part II
Current Status of and Reform Proposals for Agriculture

5
Zimbabwe

Tobias Takavarasha

Zimbabwe has had little experience with trade liberalization and market reform. There is, therefore, an urgent need to build a capacity for policy analysis to monitor and better understand the evolving liberalized market environment and provide feedback to policymakers on the emerging constraints and opportunities facing producers, consumers, and other economic agents.

Zimbabwe made impressive achievements immediately after independence, with rapid economic growth, resettlement of under-utilized land, and a sharp increase in smallholder agricultural production. The independent government inherited a highly centralized, heavily regulated economic structure with widespread state involvement in the economy. Government borrowed abroad to invest in postwar reconstruction, expanded the civil service, imposed a high minimum wage, and offered high nominal farm prices to improve agricultural incomes and production.

The system of tight import control established under the Unilateral Declaration of Independence (UDI) was maintained after independence. This, coupled with rapidly growing domestic demand, caused the real exchange rate to depreciate sharply while the nominal rate remained static. The foreign-exchange shortage worsened, private investment failed to materialize, and bottlenecks in the economy led to sluggish, stop-and-go growth. There was a deep recession in 1982-84 after the postindependence boom (Table 5.1). Some recovery occurred in 1985 in response to a good agricultural season, followed by recession in 1986/87 and recovery again in 1988-90.

These cycles are triggered primarily by rainfall, reflecting the importance of agriculture in domestic demand, export performance, and overall economic growth.[1] For the decade as a whole, real per capita

---

[1] In 1981, agriculture accounted for 11 percent of GDP (industry 43 percent and services 46 percent) and 40 percent of total merchandise exports. About half of the manufacturing sector relies on agriculture for inputs, and the agricultural sector accounts for approximately 70 percent of informal employment in the rural sector and 25 percent of formal employment.

Table 5.1—Growth rates and inflation rates, 1980-90

Rate	1980	1981	1982	1983	1984	1985	1986	1987	1988	1989	1990
						(percent)					
Real GDP growth rate (per capita)	10.6 (7.8)	12.5 (9.7)	2.6 (-0.2)	1.6 (-1.2)	-1.9 (-4.7)	6.8 (4.0)	2.6 (-0.2)	-1.5 (-4.0)	7.0 (4.2)	5.5 (1.7)	2.1 (n.a.)
Inflation rate (CPI)	4.4	13.1	10.7	23.1	20.2	8.5	14.3	12.5	7.1	15.0	22.0

Source: D. Jansen and K. Muir, "Trade, Exchange Rate Policy and Agriculture" (paper prepared for the Conference on Zimbabwe's Agricultural Revolution: Implications for Southern Africa, Victoria Falls Hotel, Zimbabwe 7-11 July, 1991).

Note: n.a. means not available.

gross domestic product (GDP) decreased by an average of 1.7 percent a year, while GDP grew by an average of 2.7 percent a year. The growth rate of agricultural output at constant prices between 1980 and 1988 was 2.2 percent, compared with a population growth rate of over 3.0 percent a year. Some of the key macroeconomic indicators for Zimbabwe are summarized in Table 5.2 for 1980 and 1990.

Blackie and Muir (1991) noted that macroeconomic policies have had a significant effect on agriculture since independence. The overvalued exchange rate, in particular, has taxed agriculture even more during the 1980s than during the UDI year (1965). Farmers producing exported commodities, particularly those who do not rely heavily on imported inputs, have been disadvantaged. This means that

Table 5.2—Key macroeconomic and structural indicators for Zimbabwe, 1980 and 1990

Indicator	1980	1990
Land area (square kilometers)	390,700.00	390,700.00
Population (millions)	7.73	9.40
Population growth rate (percent)	2.88	2.87
GDP per capita (Z$)	468.00	1,386.00
GDP growth rate (percent)	10.60	4.20
Merchandise imports (Z$ million)	860.50	4,463.91
Merchandise exports (Z$ million)	928.90	3,659.61
Current account deficit (Z$ million)	156.70	176.90
Exchange rate (US$/Z$)	1.56	0.247
Inflation (percent)	4.40	22.00
Total government spending (Z$ million)	1,225.90	6,937.70
Employment (millions)	1.01	1.18
Terms of trade	100.00	106.00
Exports/GDP (percent)	26.42	30.70
Imports/GDP (percent)	23.52	28.70
Commercial lending rates (percent)	7.50	12.00
Total money supply (Z$ million)	951.90	3,366.00

Source: Internal documents of Zimbabwe, Ministry of Lands, Agriculture and Rural Resettlement, Harare.

there has probably been a negative effect on both growth and equity, since the policies have effectively reduced employment[2] by encouraging the more capital-intensive commodities (soybeans and wheat) and more directly impacting on equity for those commodities produced by the small-scale subsector (cotton and groundnuts). All of agriculture has, however, been taxed by exchange-rate policy since independence (Jansen and Muir 1991).

Zimbabwe's poor macroeconomic performance in terms of both output and employment is attributed to a number of historical and continuing factors; among them are administrative regulations and controls that prevent productive sectors from responding to changes. The overvalued exchange rates induce an administrative allocation of foreign exchange for imports, with all the inefficiencies and distortions that are unavoidable with such a system. Exporters have little incentive to expand their operations, while their capacity to export is severely constrained by shortages of essential inputs. In addition, the foreign-currency allocation system has resulted in severe barriers to entry and highly inefficient industries.

In 1990 the budget deficit was 9 percent of GDP, total government expenditure amounted to 49 percent of GDP, while total government revenues stood at 40 percent of GDP. This level of deficit is similar to that of the previous government in the late 1970s, when a large proportion of the budget was directed toward fighting against the independence movement.

Government expenditure in the 1980s has remained high for defense and increased markedly for education and to a lesser degree for health. Parastatal deficits in recent years have grown substantially to over 40 percent of the government deficit in 1988 and over 50 percent in some years. The Agricultural Marketing Board deficits account for between 16 percent (1985) and 51 percent (1986) of the government deficit and between 2.7 percent and 5.8 percent of total government spending.[3]

In conclusion, the Zimbabwean economy since UDI has been plagued with stringent regulations, particularly in relation to trade,

---

[2] Permanent employment in the large commercial sectors stood at 228,511 in 1973 but had fallen to 136,860 in 1989. National unemployment reached 26 percent in 1989. It is estimated that there are 200,000-300,000 school leavers each year against some 20,000-30,000 new jobs created in the formal sector (Central Statistics Office 1990).

[3] These subsidies were highly regressive, as they were almost entirely consumer subsidies mainly applicable to urban areas.

agricultural marketing and pricing, general price controls, labor legislation, controls on investments, and critical input shortages, which have all contributed to inefficient industries, sluggish growth, and excessive government deficits.

NEW ECONOMIC REFORM PROGRAM, 1991-95

Growing recognition of the deleterious effects of existing economic policies was one of several influences culminating in the 1987 government decision to liberalize the economy. The shift in policies had been apparent in a number of announcements since 1988, including the establishment of a one-stop investment agency and a willingness to enter into a multilateral investment-guarantee agreement. Formal recognition of the need for a change of policy direction was first signaled in the July 1990 budget speech. A second major policy pronouncement was made in October 1990, followed shortly by the announcement that a structural adjustment program would be drawn up with assistance from the World Bank and the International Monetary Fund.

The government's Framework for Economic Reform (Zimbabwe 1991), sets out detailed measures for dealing with some of the problems outlined above, over a five-year period to 1995 (Figure 5.1). In relation to agricultural prices and marketing, the document states that "The Government is studying the modification of pricing and marketing arrangements for cotton, dairy products, meat, coffee and small grains to eliminate subsidies and allow progressive development of private marketing channels. Regional variations in prices and greater participation by private traders in marketing are goals which will be considered as part of the medium term strategy of deregulation and rationalizing the operation of the GMB" (Zimbabwe 1991, 14). The broad issues to be addressed for agriculture in the 1990s include identifying sources of future growth, increasing the efficiency with which existing resources are used, improving the equity of resource allocation within the sector, and devising policy changes and investment strategies that will foster growth. These decisions must be accomplished in a context of improved equity and conservation of the environment.

The subject of Zimbabwe's trade policy and foreign-exchange allocation system has been well documented in Jansen and Muir 1991, Takavarasha 1990, and Masters 1990. Exchange-rate overvaluation for most of the 1980s and the early 1990s is estimated at around 50 percent. A key factor in Zimbabwe's structural adjustment program is trade liberalization. Trade liberalization refers to the process of relaxing

Figure 5.1—Economic reform program in perspective

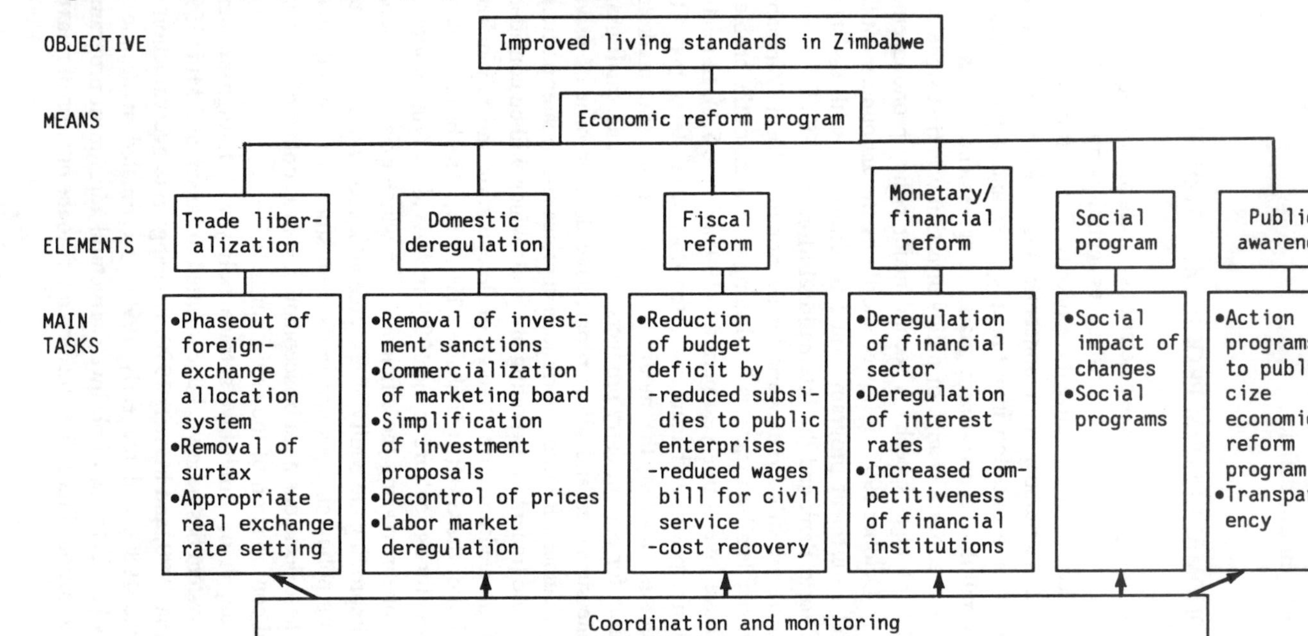

Source: Zimbabwe, Ministry of Finance, Economic Planning and Development, *A Framework for Economic Reform (1991-95)* (Harare: Government Printers, 1991).

measures affecting external trade policy, exchange rate management, and allocation of foreign exchange.

Since the 1950s, many developing countries have chosen to follow industrialization strategies based on import substitution. Quantitative restrictions, import licensing, and a complex network of controls, including the direct allocation of foreign currency by category of commodity, by type of domestic use, by source of foreign currency, and sometimes even by individual enterprise emerged as a result of this domestically oriented trade policy.

The consequence of these policies was a significantly overvalued currency, control of monetary transactions to prevent capital flight, bureaucratic allocation of foreign exchange, increased foreign borrowing, and heightened intervention by government in local capital markets (Langhammer 1987).

In Zimbabwe, protectionist policies were initiated in 1965 at the start of UDI, when economic sanctions were imposed. This system continued after independence, resulting in heavy protection of certain industrial and commercial concerns. As a result of sanctions, exports contracted and reduced the capacity of the country to earn additional foreign exchange. The available foreign exchange had to be rationed; all imports required a license granted on the basis of foreign-exchange allocation. The allocation system was based on past utilization, thus restricting new importers. The lack of both local and external competition caused by the allocation system resulted in inefficient industries and high input costs for the agricultural sector.

As far as the exchange rate is concerned, the government has pursued a policy of managed depreciation of the Zimbabwean dollar for the past 10 years. However, the demand for foreign currency continued to exceed supply throughout the period. Between May and August 1991, the exchange rate was effectively devalued by more than 30 percent. Foreign exchange, however, remained scarce, and the black market continued to operate.

An overvalued exchanged rate has the effect of implicitly taxing those sectors that are net earners of foreign exchange, in particular agriculture,[4] mining, and tourism, while a corresponding implicit subsidy is given to net consumers of foreign exchange, including the majority of manufacturing enterprises and the public sector.

---

[4] Agricultural exports in 1987 totaled Z$932 million against an estimated Z$100 million worth of imported inputs used by the agricultural sector; by 1991, exports had grown to Z$2.3 billion, while imported inputs were still severely restricted.

Historically, the emphasis in agricultural policy in Zimbabwe has been on promoting national self-sufficiency in food and raw materials, even where comparative advantage would have favored imports (for example, wheat). At the same time, the importance of agriculture in generating export revenues was acknowledged, given the limited capacity of the industrial sector to earn foreign exchange.

A number of programs have been introduced since 1983 to partially offset the foreign-exchange shortages for imports and the effects on exports of maintaining an overvalued exchange rate, and to reestablish a dynamic export sector and generate foreign exchange. These programs include the Export Revolving Fund, the Export Promotion Program, the Export Retention Scheme, commodity-specific facilities, and numerous donor-funded commodity-import programs.

Under the Export Retention Scheme, agricultural producers are entitled to retain 5-7 percent of the foreign currency earned from their crop and livestock exports, which they can use to import capital goods and essential inputs. Already it has been observed that many farmers will be induced to grow more exportable commodities at the expense of food crops. Overall, it is argued that the gains from increased export production in terms of foreign currency, higher incomes, and employment will outweigh the reduction in the production of food crops in the commercial sector. The loss in maize area, however, was already apparent even before these schemes were in place. The intention is that these export-incentive schemes will gradually be phased out as more imported goods are placed on Open General Import License (OGIL) and as the exchange-rate distortions diminish.

It would appear that the major developments in the area of trade liberalization are devaluation of the Zimbabwean dollar and placing of selected import goods on OGIL with their prices decontrolled. However, the import licensing system is still in place, implying that the barriers to entry remain high. Notable progress has been made as far as exchange-rate policy is concerned, and there are positive signs that the exporting sectors will expand, although constraints still remain in terms of the speed with which the export sectors can adjust. Their response is hampered by shortages of inputs to manufacture export products. The historic foreign-exchange allocation system is still in place, but export-promotion schemes have opened new opportunities for providing foreign exchange to exporters. Much analytical work is needed to fully understand the interaction between exchange-rate policy, tariffs, and the movement toward unrestricted foreign-exchange availability.

MARKET AND PRICE-RELATED REFORMS IN AGRICULTURE

Controlled marketing and pricing of agricultural products was established more than 50 years ago to answer a real economic, political, and social need, especially for the main food staples. The uncertainty of food production due to weather variability and the difficulties of trade in the region due to transport and foreign-exchange problems are such that governments have been reluctant to rely on private trading to supply the market for major staples (Blackie and Muir 1991). These policies appear to have had negative effects on production efficiency, employment, and possibly on income distribution, although this needs to be further analyzed. For example, a policy of controlled marketing and administered prices based, among other things, on average costs of production in a situation where there are major variations in farm sizes and resource endowments has distorted the effects of the price mechanism in its role of ensuring optimum allocation of resources and the orientation of production to meet the needs of the market place.

Public sector intervention is extensive in maize marketing in all countries in the southern African region. Analyses of the welfare effects of various stabilization schemes indicate that price stabilization can have net positive welfare implications where income and price elasticities of demand are low and where price fluctuations are due to random shifts in supply. This is particularly the case in Zimbabwe, where total maize production is highly influenced by annual variation in rainfall (Muir 1984).

It has, however, been argued that many of the problems in agriculture could be overcome by a true liberalization of agricultural prices and markets. A major thrust of many structural adjustment programs is the need to change domestic agricultural policies in order to promote production efficiency. This is because agricultural policies in many developing countries are widely believed to be biased against domestic production of food crops in favor of urban consumers, and often also against export cash crops in favor of raising government revenue (although intervention in the food marketing process usually results in a financial loss for state agencies). Therefore, reform programs have often involved eliminating food subsidies, reducing government intervention in agriculture, liberalizing agricultural trade, and reforming supportive macroeconomic policies in order to provide incentives for farmers. This is exactly what Zimbabwe's reform program entails.

Koester (Chapter 3) argues that liberalization of the domestic market is needed to support moves toward trade liberalization because such liberalization on its own cannot be a substitute for appropriate domestic economic policies. Agricultural marketing in Zimbabwe has been subject to a high degree of official regulation and administration. For a number of commodities, the government fixes producer, wholesale, and retail prices and controls the physical trading of a large proportion of these commodities through four major parastatals.[5]

As is the case with the allocation of foreign exchange for the importation of inputs, a wide variety of regulations and licensing systems govern the activities of grain producers, traders, transporters, and processors—frequently with negative effects on their operations. At independence, the government inherited a marketing system characterized by massive differences in the level of services available to the commercial and smallholder farmers. A major objective of policy during the 1980s was to improve the level of marketing services to the smallholder sector. Between 1980 and 1990 the number of Grain Marketing Board intake depots rose from 38 to 70, with all the new depots located in communal areas.

The present structure of agricultural marketing has channeled most surplus maize to the formal public marketing system and has hindered the flow of maize from surplus to deficit rural areas, thus contributing to food insecurity and sharp price differentials. The controlled marketing system has hindered the development of a low-cost and reliable marketing system in the rural areas, which in turn has militated against the diversification of cropping patterns into high-value crops in these areas (Takavarasha 1991).

There is considerable evidence that the involvement of the state in primary agricultural marketing has placed an enormous financial burden on the economies of several African states. Blackie and Muir (1991) explore the interrelations between agricultural prices and market systems, and show that farmers in Africa receive a smaller proportion of the price paid by final consumers of foodgrains than do farmers in Asia, with some 27 percent of the difference in marketing margin explained by the transaction costs of public marketing. Table 5.3 shows the budgetary consequences of the controlled marketing system in Zimbabwe and the targeted reductions over the reform period.

---

[5] The Grain Marketing Board handles maize, wheat, sorghum, groundnuts, sunflower seed, soybeans, coffee, and millets; the Cotton Marketing Board, seed cotton and cotton lint; the Dairy Marketing Board, raw milk and dairy products; and the Cold Storage Commission, cattle, beef, and beef products.

Table 5.3—Current and planned subsidies to marketing boards

Board	1987	1988	1989	1990	1991[a]	1992[a]	1993[a]
				(Z$ million)			
Grain Marketing Board	77.4	64.3	41.2	59.2	29.6	17.8	11.8
Dairy Marketing Board	49.3	51.3	52.3	59.8	29.9	17.9	12.0
Cotton Marketing Board	53.9	35.4	26.1	15.2	7.6	4.6	3.0
Cold Storage Commission	28.7	37.2	50.4	32.5	16.3	9.8	6.5
Total	209.3	188.2	170.0	166.7	83.4	50.1	33.3

Source: Internal reports of Zimbabwe, Ministry of Lands, Agriculture and Rural Resettlement, Harare; and Zimbabwe, Ministry of Finance, Economic Development and Planning, *A Framework for Economic Reform (1991-95)* (Harare: Government Printer, 1991).

Note: These figures also refer to marketing board losses. Planned subsidies are being revised downward in some boards and upward in others in view of the recently changed environment, particularly for interest rates and currency devaluation.

[a] Planned subsidies.

In 1991, maize producers in Zimbabwe were receiving Z$270 per metric ton from the Grain Marketing Board, while the board was receiving Z$320 per ton from millers, who in turn were receiving Z$800 per ton from consumers of maize meal. It has been suggested that marketing inefficiencies in Zimbabwe may also be attributed to the heavily protected processors (see Chapter 9).

The need to reform the parastatal marketing system has already been accepted by the government. It is now a question of how and when. Child, Muir, and Blackie (1985) have long suggested that a partially decontrolled market, with the government holding buffer stocks and acting as a buyer and seller of last resort, would provide a possible alternative. Agricultural pricing and marketing is seen as an integral part of the general move toward liberalized economic management in Zimbabwe. The pricing and marketing arrangements have already begun to be modified, with the understanding that an active role will be maintained by the public sector, while conditions will progressively be created for a more active private sector and greater scope for market mechanisms to determine resource-allocation decisions. Any program for maize market liberalization, however, should be carefully designed and phased to minimize adverse outcomes. Phased or partial decontrol appears to be necessary, and there are strong arguments for a continuing public-sector role in the maize market for reasons of food security and equity.

The institutional reforms should be accompanied by pricing policy reforms, institutional restructuring, and investment in infrastructure to accelerate the emergence of an efficient, low-cost, private grain-trading sector. The need to promote rural trade through provision of transport, credit and training, and market information systems is crucial. It is already apparent that the speed with which the private grain-trading sector has responded to new opportunities is disappointing.

One change already implemented in all four parastatals is the appointment in June 1991 of independent boards of directors to direct the activities, management, efficiency, and commercialization of the parastatals. The issue of greater autonomy vis-à-vis government intervention is one that is still unresolved, given that parastatals are statutory bodies set by Act of Parliament to perform commercial and developmental roles on behalf of the government. This matter can be resolved only through more dialogue and as both sides gain confidence in the changes being implemented, and as the commercial and development roles are separated. Key areas of contention are price determination, capital investments, export policy, and wage and salary policy.

PRICE REFORM

Agricultural policy in Zimbabwe has been highly interventionist, with the producer and consumer prices for most major products being administratively determined on the basis of cost of production-related criteria.

The price of maize lies at the heart of overall agricultural pricing policy because of its weight in consumer expenditure patterns and the high proportion of land and other resources devoted to maize production. A major conclusion emerging from a study of pricing policy in Zimbabwe since 1980 (Takavarasha 1991) is that pricing policy has failed to sustain opportunities for raising agricultural productivity and output and that controlled prices have failed to keep pace with inflation. Although pricing policy in Zimbabwe during the 1980s did not produce the degree of distortion that was common in other countries in Africa, the system has been very costly to maintain in terms of high transport costs, high budgetary transfers to marketing boards, and declining employment and incomes, particularly for small-scale farmers.

With the exception of maize, prices for controlled crops grown primarily in the commercial sector, such as wheat, barley, and soybeans, have either kept pace with inflation or declined only marginally in real terms. Real prices of those commodities grown widely by peasant farmers have fallen substantially. This in part reflects the concern of the government for consumer prices of basic foodstuffs, especially maize (the principal peasant crop). The decline in sorghum and millet prices is, however, market-related, as there is limited demand for these products.

The government has begun to decontrol some agricultural prices and align them with market forces and import/export parities. Regional and seasonal differentiation of certain prices is being advocated to create a more competitive multichannel marketing system that will better serve the needs of farmers, private traders, and consumers. Some of the current proposed changes in the pricing regime for major controlled commodities include the following:

- A move is needed from a system of uniform (panseasonal and panterritorial) postplanting prices (based, among other things, on cost-of-production estimates and administered through a single-channel controlled marketing system) to a more liberalized system in which prices are determined on the basis of market forces, reflecting regional and seasonal variations in supply and demand. Relaxation of some of the market regulations is expected to create a multichannel marketing system in which the private sector plays an increasingly important role.
- There will be a greater need for market information systems to monitor price trends and marketing margins for private traders. At present, most of the data used in price analysis relies on gazetted prices and the published annual reports and accounts of statutory marketing boards, which have generally been easy to find. It will be necessary to monitor and measure prices paid and quantities purchased in the informal market.
- For export crops, market information systems will be required for international price movements, changes in demand caused by technological improvements, the effects of changes in the exchange rate, and handling and transport costs necessary for calculating parity prices and supplementary payments for producers of exportable commodities.
- Market reform and liberalization will entail greater opportunities for rural traders to invest in storage, transport, processing, input supply, and agro-industries. This will require specialized training and credit facilities, development of infrastructure (roads, telephones, power, dams), and supportive policy measures such as tax incentives and foreign-exchange retention schemes for small traders, together with measures to attract established investors into

rural areas.

- There are different policy issues facing each specific commodity. Maize will remain the most strategic basic foodstuff, and it will be important for policy to sustain a reliable production base in Natural Region 2.[6] The development of a variable two-tier pricing system to build strategic food reserves and take advantage of regional exports will need to be considered further. Zimbabwe is well placed to supply seed and grain in the southern African region, where maize is the major staple. Maize production and distribution in the communal areas will benefit from improved activity by private traders. The Grain Marketing Board can complement these activities by establishing distribution depots at strategic points in grain-deficit areas as well as by providing a floor price for surplus produce. The industrial use of yellow maize for snacks, breakfast foods, starch, and stockfeed will need to be fully exploited. Pricing policy can play a role in promoting this objective.
- Because wheat is a winter crop it has significant advantages in promoting investment in water storage, irrigation, and research, which are basic requirements for food security. This suggests that the cost of producing wheat will remain relatively high.
- Pricing policy in relation to small grains has been concerned with the need to encourage communal and other small-scale producers to grow *rapoko* and *mhunga* (millets) and red and white sorghum for their own immediate consumption. These crops can play a significant role in keeping the cost of basic food low, provided varieties and milling characteristics are improved. The deregulation of the marketing system should facilitate the participation of private companies in developing these crops, but only if accompanied by product development that encourages demand.
- The market for oilseed crops (soybeans, sunflowers, groundnuts), and beef, dairy, and horticultural products remains strong within Zimbabwe and on regional markets. If these markets were completely liberalized, there would be major opportunities for growth in the supply and demand for these commodities. However, if they remain controlled, pricing policy should change from a fixed-price system to one that reflects net market realizations, so that the benefits of higher prices and signals regarding quality can be passed on to producers in the form of supplementary payments.

---

[6] Zimbabwe is divided into five natural regions that are determined by the amount and regularity of rainfall.

- In the case of cotton, steps have already been taken to replace administered prices with market-determined prices, including supplementary payments. Small-scale producers appear to have a comparative advantage in the production of cotton, which has not been fully exploited due to transport and marketing constraints. Domestic lint prices will be brought into line with export market values, thus eliminating the controversial subsidization of local, multinational textile manufacturers by cotton producers.

It is clear that the price mechanism, even under different pricing arrangements from the existing controlled system, will continue to play a critical role in increasing output, integrating small producers into the market economy, and enabling the agricultural sector to meet the requirements of the economic reform program.

CONCLUSION

Zimbabwe's economic reform program has started, by way of both stated intentions and actual implementation. The performance of the economy in terms of income growth, employment-creating exports, investment, and agricultural output in the 10 years after independence was such that a reform of the economy was needed. It is too early to make any judgments on the progress; the fact that some bold statements have been made in an economy that has seen 25 years of regulations and controls in trade, domestic marketing, prices, investment, and employment is very encouraging.

Many impediments are likely to exist, particularly where food commodities are concerned. The amount of flexibility that has been given to the Cotton Marketing Board is not comparable to that given to the Grain Marketing Board for maize, which is the staple. The inability of the liberalized sectors to respond quickly by providing more goods or more foreign currency, and for the private sector to complement the parastatal reforms, clearly reflects the legacy of too many years of controls. In terms of equity, it would appear that the benefits of liberalization are accruing to traditional exporters and to monopolies, who continue to have protected access to foreign exchange, or who can charge excessive prices. It will be some time before barriers to entry are eliminated and new entrants can compete and force prices down.

The rapid inflation that so far has been clearly demand-driven now appears, in addition, to be cost-pushed by the effects of devaluation on the imported raw materials needed for the manufacture of scarce goods. On the positive side, many exporting companies have seen a sharp increase in their revenue and capacity to import (because of devaluation and export retention schemes). Already there has been an increase in employment in tobacco, horticulture, and other exporting sectors.

Total foreign-exchange earnings from agricultural exports for 1991 are estimated at Z$2,296 million (an increase of 55 percent over the revised estimate of Z$1,492 million for 1990). The major contribution to this total is again from tobacco; total flue-cured plus burley tobacco exports for the current year are estimated at Z$1,658 million. The next-largest contribution is from cotton lint (Z$197 million), followed by horticulture (Z$125 million). The preliminary estimate for 1992 is that exports will increase by a further 45 percent, but this is likely to be an underestimate. As the full benefits of devaluation work through, it appears from internal reports of the Ministry of Lands, Agriculture and Rural Settlement that agricultural exports will generate closer to Z$4,000 million. While benefits are high for those with access to foreign currency and import substitution, domestically oriented industries still under price control and the low-income groups are beginning to feel the adverse effects of adjustment (for example, declining market shares and production capacity, declining purchasing power and high rates of inflation).

The conclusion drawn is that rapid monitoring and analysis of the effects of the Economic Structural Adjustment Program (ESAP) on various aspects of the economy need to be put in place so that the adverse effects can be corrected and the positive effects of adjustment can be capitalized on. Work has already started in the Ministry of Lands, Agriculture and Rural Settlement to set up a data base for monitoring the effects of ESAP on agriculture, using the policy analysis matrix methodology (Takavarasha, forthcoming). The focus will be on (1) trends in output and shifts in production between crops and the effects of these trends and shifts on employment and income distribution; (2) trends in price levels and their effects on food security; and (3) trends in input usage, in particular, fertilizer and machinery. The effects of exchange-rate and interest-rate policy on agriculture will also be studied.

It is evident that there are major problems of evaluation and analysis that will need to be tackled. The policy problems are very large; it would be a mistake to assume that movement toward a more market-determined system will lessen the need for well-formulated policy decisions and the requirements for good policy analysis. There is an enormous task ahead, which will make very large demands in terms of policy evaluation and data collection.

REFERENCES

Blackie, M.J., and K. Muir. 1991. Economics of the agricultural production revolution. Paper prepared for the Conference on Zimbabwe's Agricultural Revolution: Implications for Southern Africa, 7-11 July, Victoria Falls Hotel, Zimbabwe.

Central Statistics Office. 1990. *Statistical yearbook.* Harare: Government Printer.

Child, B., K. Muir, and M. J. Blackie. 1985. An improved maize marketing system for African countries: The case of Zimbabwe. *Food Policy* 10 (4): 365-371.

Jansen, D., and K. Muir. 1991. Trade, exchange rate policy and agriculture. Paper prepared for the Conference on Zimbabwe's Agricultural Revolution: Implications for Southern Africa, 7-11 July, Victoria Falls Hotel, Zimbabwe.

Langhammer, R. J. 1987. Implications of foreign trade liberalization in developing countries. *Indian Journal of Economics* 68, Part 2 (October).

Masters, W. 1990. *The value of foreign exchange in Zimbabwe: Concepts and estimates.* Working Paper AEE 2/90. Harare: University of Zimbabwe, Department of Agricultural Economics and Extension.

Muir, K. 1984. Crop price and wage policy in the light of Zimbabwe's developing goals. Ph.D. diss., University of Zimbabwe, Department of Land Management.

Takavarasha, T. 1990. The role of agriculture in the process of structural adjustment in Zimbabwe. Paper prepared for the Ministry of Land, Agriculture and Rural Resettlement, Harare.

_____. 1991. An analysis of Zimbabwe's agricultural price policy since independence. Paper prepared for the Conference on Zimbabwe's Agricultural Revolution: Implications for Southern Africa, 7-11 July, Victoria Falls Hotel, Zimbabwe.

_____. Forthcoming. Institutionalizing the analysis of producer price policy in Zimbabwe using the policy analysis matrix methodology. Ph.D. Research Proposal, University of Zimbabwe, Department of Agricultural Economics and Extension, Harare.

Zimbabwe, Ministry of Finance, Economic Development and Planning. 1991. *A framework for economic reform (1991-95).* Harare: Government Printer.

6
Zambia

Julius J. Shawa

Zambia is a landlocked country with a current population of about 8 million and an average annual population growth rate of about 3.3 percent. The country is sparsely populated but highly urbanized (slightly more than 50 percent of the population lives in urban areas) despite abundant underutilized land with adequate rainfall patterns over most of the country.[1]

Zambia's economy has declined since the mid-1970s, most markedly in recent years. The economy is heavily dependent on copper export earnings, which account for 80-90 percent of total earnings. The fall in copper prices that started in the mid-1970s, the excessive imports for both consumption and production, and a poorly developed agricultural sector have contributed to the decline in the economy. The government has therefore been relying on external borrowings to pay for imports. There has also been heavy dependence on donor assistance to support certain projects and programs in some sectors including agriculture. The country's external debt is currently estimated to be about US$6.8 billion. At present the conflicting demands for debt-servicing and for financing of essential imports are posing severe constraints on the ailing economy.

Real gross domestic product (GDP) stagnated during 1980-87, averaging only 0.2 percent growth per year. The net result of the poor performance of the economy has been a marked decline in the people's standard of living. The World Bank has estimated that real per capita gross national product (GNP) in 1989 was about US$290—only half of the 1980 level. Economic performance has also been adversely affected by strong public intervention in the economy and controls on production, pricing, marketing, proçessing, and other economic activities.

Since the late 1970s the country has embarked on various structural

---

This chapter was written a month after the election of the new government in Zambia and reflects only the reforms undertaken by the Kaunda regime.

[1] Zambia cultivates only 2.5 million hectares of the 42 million hectares of potentially arable land.

adjustment programs aimed at resuscitating the ailing economy. Most of these programs have had the full endorsement and support of the International Monetary Fund (IMF) and World Bank and other bilateral donor agencies. Various structural adjustment programs have been abandoned or modified according to political imperatives. The current program was initiated in 1989, when the government adopted a market-oriented adjustment strategy. This strategy aims to restore economic growth through financial stabilization and to diversify production and exports from copper to agriculture and other high-potential sectors such as tourism.

The country has the potential to grow a variety of crops and to rear more livestock. Maize is the predominant crop and a staple food. It occupies about 60-70 percent of the cultivated land. Other important crops grown include cassava, sorghum, millet, cotton, tobacco, sugar, rice, mixed beans, sunflower, groundnuts, and soybeans. Agricultural production is carried out by four categories of farmers: small-scale farmers (peasants); emergent (small commercial) farmers; commercial farmers; and institutional farmers.

The country can be divided into four agroecological zones: the northern, high-rainfall zone that occupies about 46 percent of the country; the central, southern, and eastern plateaus, which constitute about 12 percent of the area; the western, semiarid lands that occupy 28 percent of the area; and the rift valley zone that accounts for the remaining 14 percent.

The average annual growth of the agricultural sector over the past two decades has been about 2.5 percent, well below the population growth of 3.3 percent. Self-sufficiency in major food crops has consequently declined. As a result, Zambia has become a net importer of food, notably edible oils, wheat, dairy products, and in some years, maize. In the early 1980s, food imports accounted for 10 percent of the total value of imports, while agricultural exports (mainly tobacco, cotton, and groundnuts) accounted for less than 2 percent of the country's export earnings.

For some time, the government's agricultural policy has been to increase agricultural production and reach self-sufficiency in the staple food crops and import-substitution crops as well as to increase production of exportable agricultural commodities. In the past, the Zambian agricultural marketing system was typified by state intervention at every stage. This was done through marketing boards and cooperatives that have a monopoly in produce marketing. In addition to the control of the agricultural market at the producer, wholesale, and retail levels, the government maintained strict control of retail prices for a wide range of goods. The granting of monopoly power to marketing boards and cooperatives effectively stifled private-sector involvement.

In line with economic structural adjustment, the government has accepted that regulation and control of agricultural marketing and prices, coupled with heavy subsidies, are unsustainable and discouraging to agricultural development. Hence, in June 1989, prices of all crops except maize were deregulated together with all other previously regulated products. Changes of policy have continued, and in September 1990 an important measure was introduced to allow farmers to sell maize directly to millers, cooperatives, and other private traders. Until that time, maize was sold only to the cooperatives and through them to the mills at government-fixed prices. Fertilizer marketing, including imports, was also opened up. Although a step in the right direction, the new measures on maize marketing have fallen short of complete liberalization because both the into-mill and retail prices of maize and maize meal remain controlled.

ECONOMIC POLICIES AND DEVELOPMENTS FROM 1985 TO 1988

In late 1985, Zambia, with the support of both the IMF and World Bank, launched an economic restructuring program that involved wide-ranging macroeconomic and structural reforms. These included the need for tighter domestic financial policies, market-determination exchange rates through the auction mechanism, and market-determined consumer prices. These measures were to be accompanied by tax, tariff, and public-enterprise reforms.

The adjustment program was seriously undermined by shortfalls in copper export earnings and by a rapid acceleration in money supply. The sharp currency depreciation weakened confidence in the auction system. For instance, at the start of the auction system in October 1985, the exchange rate was 5 Zambian kwacha (K) to 1 U.S. dollar, and by the end of the system in March 1987 it was K15/US$1, a 150 percent depreciation over an 18-month period. Economic growth dropped to below 1 percent, inflation rose to over 50 percent, and both the budget and external current deficits widened. Foreign-exchange reserves fell sharply. After food riots in 1987, which were sparked by increased maize-meal prices, the government broke off from the IMF-sponsored economic adjustment program. This move undermined the vigorous restructuring program, and black market exchange rates subsequently depreciated the kwacha even more rapidly than the auction did.

In May 1987, the government introduced its own New Economic Recovery Program (NERP), whose theme was "growth from own resources." The strategy limited external debt-servicing, fixed interest rates, reintroduced price controls, returned to centralized control of all foreign currency, and reduced the rate of increases in the money supply from 60 percent in 1987 to 40 percent in 1988.

Economic performance in the following year (1988) was strong in certain respects. A growth of 6.7 percent was recorded, compared with 2.2 percent in 1987. The major impetus of this growth came from the agricultural and manufacturing sectors, where output increased by 21 percent and 15 percent, respectively. The manufacturing sector benefited from increased foreign-exchange allocations made possible by high copper prices and higher capacity utilization rather than new investment. Agricultural output, especially maize, was helped by increased producer prices, improved credit and extension services, timely payment, and good rainfall.

Despite achievements in some sectors, other sectors of the economy, notably mining, construction, and transport, registered negative or zero growth rates, and there was a decline in formal sector employment. During 1988, inflation rose to 64 percent. The primary cause of this acceleration trend was excessive monetary expansion stemming from the budget deficit, strong credit expansion to the agricultural sector, and improvement in international reserves.

The country continued to experience a chronic shortage of foreign exchange that deprived the economy of essential inputs and basic consumption goods and constrained the servicing of foreign debt, which in turn stifled access to new external credits. A number of remedial economic measures were taken in late 1988 and early 1989 to address some of the problems identified above. These included devaluating the kwacha by 25 percent in November 1988 and pegging it to the special drawing rights (SDR), raising the minimum reserve requirements by 5 percent during the same period, and reducing the budgetary cost of maize subsidies by increasing maize meal prices in January 1989 and introducing the coupon scheme.

Recent agricultural and economywide policy reforms in Zambia can be traced to the Policy Framework Paper (PFP) of August 1989, revised in February 1991 in the Fourth National Development Plan (1989-93). The immediate macroeconomic goal is to bring down inflation and to create a stable economic climate for growth and diversification.

The PFP spells out the first priority in the economic adjustment strategy as the stabilization of the economy by eliminating market distortions. Structural and sectoral policies in the PFP are geared to shifting resources to more labor-intensive activities, including small-holder agriculture and small-scale industry. The PFP represents the first phase in Zambia's medium- and long-term adjustment program. Among the basic policies and reforms needed to restructure and raise the efficiency of the economy are

- Developing other sources of export earnings (nontraditional exports) in view of the declining copper market and exhaustion of the ore reserves;
- Adjusting the exchange rate of the kwacha to attain a rate that clears the market. This will greatly assist in attaining rapid growth of nontraditional exports;
- Raising the saving and investment ratios. The highest priority for investment is to rehabilitate and refurbish existing productive capacity and infrastructure;
- Developing the country's manpower base in line with the needs of the economy, reducing high population and labor growth rates, and protecting the environment;
- Decontrolling prices of all commodities;
- Raising interest rates in stages, with the aim of having positive real rates by 1991;
- Adopting and consolidating other exchange and trade policies, such as export retention, and decentralizing export licensing;
- Broadening the parastatal and public expenditure reform efforts in order to raise efficiency and increase the productivity of resource use by the public sector;
- Introducing tariff reform, such as keeping protective tariffs to a minimum; and
- Encouraging participation of the private sector in the economy in order to promote greater competition as a means of improving economic efficiency.

The PFP sets specific targets and dates for achieving some of these economic objectives. The revised PFP of February 1991 goes even further by bringing forward the dates for achieving these targets. (Table 6.1).

A Public Investment Program (PIP) complementing the PFP was drawn up in 1990 and covers the period 1990-93. The PIP is a high-priority set of projects drawn from the Fourth National Development Plan. It covers those programs that can be implemented with the resources available, relying on increased donor support and foreign

Table 6.1—Timetable of stabilization and structural measures for the agricultural sector, 1990-93

Measures	Timing of Measures
Permit private trading in maize and fertilizer	Implemented September 1990
Conduct survey of maize stocks in storage facilities to ensure adequate supplies for 1991/92 marketing season	1990 to early 1991
Revise rules for and implement revised floor pricing system with seasonal and geographical variations[a]	Early 1991
Establish clear rules for operation and financing of maize strategic reserves	Early 1991
Decontrol domestic price of maize and maize meal production[a]	Mid-1991
Eliminate transport subsidies for maize[a]	1993
Phase out all subsidies on fertilizer and decontrol fertilizer prices[a]	1991
Implement hammer mill program and encourage other private milling to stimulate competition in domestic milling industry	1990-93
Establish priority expenditure program for the agricultural sector, emphasizing extension, research, veterinary services, and rural transport	1990-93

Source: Zambia, *New Economic Recovery Program: Economic and Financial Policy Framework, 1991-93* (Lusaka: Ministry of Finance and National Commission for Development Planning, 1991).

[a] Priority area for early action.

funding. The PIP gives priority to rehabilitation and maintenance of existing assets and facilities, with particular emphasis on completion of ongoing projects. No new projects will be undertaken outside of the framework of the PIP.

Complementing the PIP is the New Investment Act, which seeks to encourage investments in Zambia by both local and foreign investors. The Investment Act of July 1991 seeks to revise the law relating to investment in Zambia. The Act favors exporters of nontraditional products or services resulting in net foreign-exchange earnings, producers of goods for use in the production of agricultural commodities or other agrorelated products for exports, businesses engaged in tourist activities that result in foreign-exchange earnings in excess of 25 percent of their gross annual earnings, import-substitution industries that use a significant proportion of local raw materials, and exporters located in rural areas.

The Act also contains a number of incentives for investors:

- Exemption from customs duties and sales tax on all machinery equipment and parts for the establishment, rehabilitation, or expansion of an enterprise;
- Exemption from tax on dividends for a period of seven years from the date of commencement of the business;
- Exemption from payment of tax on income for companies for a period of three years from the date of commencement of business;
- Exemption from the payment of selective employment tax for a period of seven years; and
- Retention of 70 percent of its gross foreign-currency earnings for the first three years and 60 percent of such earnings for the remaining period of validity of the investment license.

In addition to these, there are other incentives for agricultural and tourist enterprises, import-substitution industries, and small-scale and village enterprises. The Act also sets investment guarantees that include protection from acquisition and settlement of disputes.

Development of a competitive agricultural sector is an essential element of the government's structural adjustment strategy as asserted in the Fourth National Development Plan. The objectives regarding this sector are to achieve and maintain self-sufficiency in food production as well as to generate a major share of the country's future export growth. Agricultural policy is also aimed at providing food (maize) security at the national and household levels, and encouraging full exploitation of the country's large potential for supplying primary commodities to agroindustry and exporting agricultural products such as coffee, cotton, tobacco, fruits and vegetables, horticultural products, and beef.

To achieve these objectives, the government will continue to restructure the agricultural marketing and pricing system. According to the PFP, structural-adjustment measures to be taken in the agricultural sector include phasing-out of handling subsidies, equilibrium pricing of all agricultural crops and commodities, increased private-sector participation in maize marketing, and removal of all constraints on exports of agricultural commodities (including maize in surplus years).

The intended effect of these various policy changes is to realize the efficiency gains from comparative advantage and to encourage production of export crops. It is expected that export production will take place in both the smallholder sector and commercial agriculture.

Since 1989 the restructuring program for the agricultural sector has involved

- Abolition of NAMBOARD, a monopoly board in crop (maize) and fertilizer marketing. Maize marketing was taken over by the cooperatives, and fertilizer imports and marketing by Nitrogen Chemicals of Zambia, while the strategic maize reserves function was transferred to the Zambia Cooperative Federation;
- Introduction of the maize-meal coupon program to families with six or more dependents and with income levels not exceeding K 20,500. This was meant to cushion the impact of increased meal prices on low-income urban consumers; and
- A move by the government to phase out maize and fertilizer subsidies as well as the fertilizer price differential subsidy.

CURRENCY STATUS AND IMPACT OF POLICY REFORMS

Macroeconomic Policies

Money Supply. After achieving a sharp decrease in money-supply growth and inflation through the first half of 1990, policy slippage, compounded by the effects of higher international oil prices during the gulf war, resulted in fiscal and monetary deterioration, inflation, and poor balance-of-payments performance during the second half of 1990. Money supply has continued to grow, partly as a result of increased government expenditure, hyperinflation and negative real interest rates in the formal sector, and low confidence in the banking sector.

Exchange and Interest Rates. The movement of exchange rates during the period 1987 through November 1991 is indicated in Table 6.2. The kwacha showed a steady depreciation against the U.S. dollar and other major currencies during this period. In February 1990 the government accelerated the pace of effective exchange-rate adjustment by introducing, as a transitional arrangement, a dual foreign-exchange market that consisted of two windows. At the first window, foreign-currency proceeds from copper exports were sold for specified purposes at the official exchange rate. The second window was fed by other foreign-exchange earnings, including nontraditional exports and external loans and grants. In April 1991 the two rates were merged into one. The country continues to face negative real interest rates with inflation running at over 100 percent (Table 6.3).

Table 6.2—Average official and market exchange rates, 1987-91

Year	Official Rate	Market Rate
	(kwacha)	
1987	8.69	n.a.
1988	8.27	10.00
1989	13.84	21.65
1990	30.59	40.82
1991 (November)	83.75	85.88

Source: Bank of Zambia, *Main Economic Indicators* (Lusaka: Bank of Zambia, 1991).
Notes: n.a. means not available. Official and market rates merged on April 24th, 1991.

Table 6.3—Real interest rates, 1985-90

Year	Central Bank Rate	Inflation Rate	Real Interest Rate[a]
		(percent)	
1985	17.50	36.83	-14.13
1986	25.00	52.51	-18.04
1987	15.00	44.48	-20.40
1988	15.00	54.23	-25.44
1989	19.00	122.41	-46.50
1990	29.00	109.60	-38.45

Source: Bank of Zambia, *Main Economic Indicators* (Lusaka: Bank of Zambia, 1991).
[a] Real interest rate = $[(1 + r)/(1 + i)] - 1$, where r = Central Bank rate, and i = inflation.

Budget Deficit. The primary budget deficit, defined as the overall deficit (on an accrual basis) excluding domestic and foreign interest and grants, was reduced from 7.1 percent of GDP in 1988 to 6.1 percent in 1989. Real GDP was broadly unchanged in 1989. During 1990 there was a large increase in government expenditure as a result of a significant increase in housing allowances that were granted to civil servants not living in government housing. These allowances were tax free and have been very costly to the budget. Apart from this, other significant expenditure overruns occurred in other areas of the budget, owing largely to inadequate control over spending by individual ministries. On the other hand, revenue performance in all tax categories improved, with tax revenue from the nonmineral sector expected to show an increase of more than 1 percentage point of GDP to 14 percent. As a consequence, despite spending overruns, it is estimated that there was a further sizable reduction in the primary deficit to 4.5 percent of GDP in 1990.

Parastatal Reform (Privatization). There are currently over 120 state-owned enterprises. The government has identified the initial seven state-owned enterprises that are up for sale to interested buyers, be they local or international. The partial privatization program is intended to span a maximum period of five years, up to 1996. The state will retain controlling shares in the enterprises.

Coupon Program. The coupon program is a targeted food-subsidy system aimed at cushioning the impact of subsidy removal or reduction and the subsequent rise of the meal price faced by the formal-sector wageworkers. Those living in the rural areas are excluded from the program on the basis that they are supposed to grow their own maize. The administration of the program has proved difficult because households for which the program was not intended (the unemployed and those in the informal sector) have benefited from it although the government intended that only the wageworkers should profit. This has led to a review of the program with the aim of restructuring and making it more effective.

Agricultural Sector Performance

It can be argued that most of the objectives for the agricultural sector have not been realized. Production of agricultural products, especially crops, fluctuated a great deal between 1980 and 1991.

The prices paid to the farmer, inefficient marketing systems, weather conditions, and structural-adjustment policies have all contributed to the fluctuation in crop production. Maize production for instance, has declined sharply over the 1989/90 and 1990/91 seasons. Marketed maize production in 1989/90 was only 5 million bags. In the 1990/91 season, slightly more than 6 million bags were marketed. Consequently, the maize shortfall in early 1992 was expected to be as much as 3 to 4 million bags.

The objective of self-sufficiency in maize in some years has been achieved at great cost to the national economy. Self-sufficiency in maize production has been achieved at the expense of other staple food crops that have comparative advantage in most of the regions in which maize has been promoted. The agricultural sector's contribution to growth in GDP averaged 2.5 percent (at constant 1977 prices) from 1980 to 1990. This contribution is low relative to the potential of the agricultural sector.

The objective of import substitution is yet to be achieved, as the country still imports about 60 percent of its wheat requirements, 50 percent of its rice, and large quantities of edible oil and barley. Agricultural export performance has not been encouraging. Agricultural export earnings as a percent of total export earnings have been negligible at 2 percent. Agricultural exports are dominated by tobacco, coffee, cotton, sugar, and horticultural crops.

Macroeconomic Policy Impact on Agriculture. One of the two basic objectives of the structural-adjustment policies is the reduction of government expenditure. Agricultural expenditure as a percentage of total government expenditure declined from 23 percent in 1980 to about 15 percent in 1987 and then to 4 percent in 1990. The reduction in government expenditure has had a negative impact on the growth of the sector.

The other important policy objective has been to control the growth in money supply (credit squeeze). This, compounded by increased interest rates and a shortage of credit, has had a negative impact on agriculture. There has also been inadequate funding for crop purchasing. This has been worsened by the high interest rates and mismanagement of agricultural marketing institutions.

Agricultural Marketing Liberalization. Another major policy reform undertaken by the government has been the liberalization of agricultural marketing for both crops and inputs, notably fertilizers. Liberalization of marketing of other crops (other than maize) has to a large extent been successful. For instance, private traders' participation has been widespread in the procurement and sale of rice, wheat, sorghum, millet, cassava, soybeans, and groundnuts. In addition, private traders have directly negotiated with farmers for producer prices. Some of these crops such as cassava, millet, and sorghum, which are commonly referred to as traditional crops, have not been handled by official marketing institutions on a large scale because of the lack of a market for these crops in urban areas. Therefore, these crops continue to be handled by the private traders. Agrochemicals, veterinary drugs, and farm equipment and implements are also dominated by the private sector, as are trade in livestock, vegetables, and fruit. Liberalization has led to increases in production of some crops, notably soybeans and wheat. Commercial farmers and, to some extent, small-scale farmers have shifted their production patterns away from controlled maize to these uncontrolled crops.

Pressure from the donor community and severe financial constraints on the government contributed to the liberalization of maize and fertilizer marketing in 1990. The liberalization of maize and fertilizer means that farmers, primary cooperative societies, and cooperative unions are allowed to sell maize directly to millers, retailers, or private traders and members of the public at floor or negotiated prices; it also means that farmers, private traders, or firms with their own foreign exchange are permitted to import fertilizers for own use or for sale in the domestic market.

After one year, a number of problems are seen in the liberalization of maize and fertilizer marketing.

- *Minimal private sector response.* Private traders are cautious about investing in unfamiliar operations that the government may recontrol in the future. Confiscation or seizure of private milling companies and some retail outlets by the state in 1986 has discouraged traders' confidence. It would appear that few private traders will participate in maize marketing, as profits are low and the future uncertain. Very few, if any, private traders are involved in fertilizer importation and distribution, partly because of the subsidy that is still being paid by the government on certain types of fertilizer, notably those used on maize. In addition, since fertilizer is a bulky product, it requires a huge outlay of funds, which private traders cannot raise, for procurement, transportation, and storage. It should be noted that major milling companies are participating in the purchase of maize, while some companies or privately owned cooperatives, such as the Zambia Tobacco Company, Zambia Farmers' Cooperative, and Agricoop, are participating in the distribution and selling of fertilizer.
- *No proper market information system.* Having an adequate market information system in place is necessary to enable traders to make sound business decisions, give farmers signals, and enable the government to make proper policy decisions.
- *Inadequate funds to purchase crops and import fertilizer.* Crop financing has been a chronic perennial problem for cooperative unions, millers, and others during each agricultural marketing season. For the 1990/91 marketing season, the government had to directly provide funds (K 3.5 billion) for the purchase of maize. As a result of huge overdrafts that the cooperative unions have had

to incur with the commercial banks, the banks have been reluctant to release any more funds to the unions. The government, as the guarantor of the overdrafts, had to step in and clear the outstanding debts, which amounted to K 1.2 billion. With regard to fertilizer importation, the government had to provide the kwacha funds to Nitrogen Chemicals of Zambia, which had not collected K 1.07 billion owed to them by the unions for the previous year's fertilizer sales. The problem of insufficient funds has been compounded by the credit squeeze coupled with high nominal interest rates that have made it difficult for marketing institutions to borrow from commercial banks.

- *Contradictory government policies*. While the government has verbally committed itself to liberalization of maize and fertilizer marketing, it has on the other hand been increasing subsidies on maize and maize meal in particular. This is expected to change with the new government.
- *Insufficient storage facilities*. Panseasonal pricing policies discourage on-farm storage by millers and traders. Currently there are no clear policies regarding the usage of storage facilities. The government has stated that it owns the storage facilities, but it is not clear which types of storage facilities the government owns and who is responsible for their maintenance. To date, no rental fees have been charged for using the facilities. Therefore, a clear-cut policy is needed on the use of the existing storage facilities.
- *Poor road infrastructure*. This poor infrastructure, especially in the maize surplus areas, hampers maize marketing and fertilizer distribution. There is a need, therefore, for the government to rehabilitate feeder roads in the high-production areas. In addition, local institutions such as district councils should be encouraged to generate adequate resources for the maintenance of feeder roads.
- *Negative effect on credit recovery*. The liberalization of agricultural marketing has had a negative effect on the recovery of credit by lending institutions, notably the three that cater to small-scale farmers.

Despite these problems, it can be deduced from the various measures and utterances that the new government is committed to total liberalization of maize and fertilizer marketing. This is so because the government knows that decontrol of maize and fertilizer prices is important not only to reduce the subsidy burden on the budget but also to improve the structure of incentives to farmers and consumers and thereby improve the efficiency of agricultural production and marketing.

CONCLUSION

The country's economic restructuring program was not realized due to the previous government's refusal to implement remedial measures agreed upon with the IMF and World Bank. For instance, the operational results on inflation, nontraditional exports, and monetary and fiscal aggregates fell very short of the established targets. Notable programs that the previous government refused to implement, in view of the elections that were held at the end of October 1991, included increases in fuel and meal prices. However, maize meal in most parts of the country was trading at prices above government-announced levels. The previous government also went against the agreement by awarding the civil service huge salary increments across the board in August 1991. These actions, together with failure by the government to settle US$20.8 million in arrears owed to the IMF and World Bank perturbed the two institutions and led to the suspension of relations with Zambia.

The country's precarious economic position has been worsened by the maize shortfall in 1991, which necessitated imports from South Africa. Many hopes are being placed on the newly elected government (October 1991), but it is impossible to be sure of its commitment.[2] Meanwhile, donors have indicated that political pluralism alone will not bring automatic financial benefits to the ailing Zambian economy unless the new government presents a workable plan for restructuring and reaches an agreement with the IMF.

---

[2] Early indications are good, and the civil service was substantially reduced within the first three months after the election.

REFERENCES

Bank of Zambia. 1991. *Main economic indicators*. Lusaka: Bank of Zambia.

Zambia. 1991. *New Economic Recovery Program: Economic and financial policy framework, 1991-93*. Lusaka: Ministry of Finance and National Commission for Development Planning.

7
Malawi

Katundu M. Mtawali

The economy of Malawi is highly dependent on agriculture. The importance of the agricultural sector is evidenced by its share in the gross domestic product (GDP) (almost 37 percent) and its share in total export earnings and total employment (over 85 percent for both). More than 90 percent of the population, which is currently estimated at over 8 million and has a growth rate of 3.5 percent, is rural based. The slow growth of the industrial or manufacturing sector has meant, and still means, that the agricultural sector will continue to shoulder the burden of providing a livelihood for a large proportion of Malawi's growing population. It is not surprising, therefore, that policy action in Malawi, both agricultural and economywide, is largely based on influencing the dynamism of the agricultural sector.

Agriculture in Malawi, as in most Sub-Saharan economies, has been characterized by a degree of dualism that has dichotomized the sector into smallholder (or peasant) and estate subsectors. This dualism is largely a reflection of colonial policy strategy, which emphasized the development of estates to produce export crops, while the smallholder sector was mainly regarded as a subsistence subsector for the indigenous people. As time went on, however, production of certain cash crops that were originally a monopoly of the estate subsector found their way into the smallholder subsector.

All production that takes place on leasehold land and in freehold tenure systems constitutes the estate subsector, while all agricultural activities that take place on customary land constitute the smallholder subsector. Some statutory agencies undertake limited agricultural activities on public land, especially for tobacco, tea, and sugar.

Another feature that distinguishes the estate subsector from the smallholder subsector is the pricing and marketing policy pursued in each subsector. Estate producers sell directly to the final markets, so prices for the commodities are largely determined by supply and demand conditions, although these markets are affected by distortions. The smallholder, on the other hand, is required to sell all his output through the government agencies.

Finally, the major source of capital for the smallholder is the government, while commercial banks are the main source of financing for the estates, since the title to land provides acceptable collateral.

POLICY ENVIRONMENT AND ECONOMIC PERFORMANCE BEFORE REFORM

The performance of the Malawi economy during the 1970s has been variously described as "impressive" (Sahn, Arulpragasam, and Merid 1990), and "among the strongest in Sub-Saharan Africa" (Kydd and Christiansen 1987). The average annual growth rate of GDP at 4.9 percent in the 1960s accelerated to 6.3 percent in the 1970s. The per capita food production during that period was maintained above the population growth rate, while most low-income countries experienced an average drop of 12 percent (Sahn, Arulpragasam, and Merid 1990). This impressive performance in the 1970s largely reflects the good performance of the agricultural sector, particularly the estate or export subsector. The question is, then, what caused this strength in the economy during the 1970s?

Malawi's bias toward an export-oriented, estate-led growth strategy is clearly manifested in the policies pursued in the agricultural sector. Incentives to the sector included a monopoly on the production of high-value crops (for example, tea, sugar, and burley and flue-cured tobaccos were confined to estates). This bias toward the estate sector in terms of policy incentives is also apparent in market arrangements for inputs and outputs. While the smallholder producer was asked to conduct his business through middle markets (Farmers Marketing Board, Agricultural Development and Marketing Corporation [ADMARC], and Press Holdings) for both inputs and outputs, the estate producer was allowed to conduct business in final markets. ADMARC made substantial profits from smallholder tobacco, averaging K 14 million per year in the 1970s (Muir 1982). This smallholder tax was used to subsidize urban maize consumers and to invest in the estate sector and agroindustries. These investments have been fraught with problems of mismanagement and have been disappointing. Policy bias toward the export sector is also revealed by the ease of access to land for production.[1]

---

[1] In 1970 there were only 229 leasehold estates, but by 1979 the number had increased to 1,105, a more than fivefold increase within a decade (Mkandawire 1990).

Recognizing the importance of competitiveness in the international markets, Malawi has pursued a policy characterized by relatively liberal exchange-rate regime and trade arrangements. Prior to independence and immediately after, Malawi's main trading partner was the United Kingdom, and Malawi pegged her currency to the pound sterling. As markets for the country's products began to diversify, its currency was pegged to a weighted average of currencies that constitute the Special Drawing Rights (SDR). Since its currency was aligned with currencies of the major trading partners, Malawi's exchange mechanism moved more or less in resonance with their exchange rates, thus ensuring competitiveness. Malawi has avoided policies of deliberate currency overvaluation to tax agriculture. Import duties on agricultural inputs were kept low throughout the 1970s. Such duties ranged from 2 to 3 percent for machinery and equipment and from 20 to 45 percent for consumer goods and petroleum products.[2] The application of quantitative restrictions was also limited.

Another important policy that provided incentives to the estate sector is the indirect taxation of smallholder output. The prices paid by the statutory agencies to smallholders during the 1970s and even during most of the 1980s were well below the final market prices realized by these boards. The funds so generated were not reinvested in smallholder agriculture but primarily in the estate subsector, with some funds actually being invested in nonagricultural activities.

The performance of saving and investment components during the 1970s was conducive to economic growth. While the 1960s saw growth in consumption exceeding growth in GDP, the 1970s were characterized by low consumption growth rates averaging 4 percent annually in real terms. During the 1970s, government emphasis switched to building the material and technical base of the economy. Restrained consumption during the 1970s facilitated increased saving and investment levels.

Still another important policy variable that has had a bearing on the success of Malawi's export-led growth is the wage policy. Malawi's agriculture is only moderately capitalized. The government, recognizing the labor-intensive nature of agriculture, deliberately pursued a policy of wage restraint that further biased incentives in favor of the estate sector. Minimum wages in nominal terms were virtually static between 1970 and 1980, and real wages actually declined. There was a modest

---

[2] The estate sector is the more mechanized of the two subsectors, and the low import tax for farm machinery reflects incentives biased toward the export subsector.

increase in real wages in the 1980s (Table 7.1).

In addition to growth-facilitating policies, there were exogenous factors that augmented the good performance of the economy, particularly the agricultural sector. The most significant of the exogenous factors was the imposition of trade sanctions on Rhodesia by the international community because of that country's unilateral declaration of independence. The trade embargo on Rhodesia partially contributed to the improved terms of trade experienced by Malawi in the 1970s. The embargo also created an exodus of producers and managers from Rhodesia into Malawi and these helped to improve the quality of tobacco management skills in Malawi.

FACTORS THAT NECESSITATED REFORM

The sound economic performance of the 1970s did not last long, however. Several factors explain the subsequent downturn, and these include both policy and nonpolicy factors. The export-led growth

Table 7.1—Statutory minimum daily wage rate, 1980-89

Year	Blantyre/Lilongwe		Zomba/Mzuzu		Other Areas[a]	
	Nominal	Real[b]	Nominal	Real[b]	Nominal	Real[b]
			(kwacha/day)			
1980	0.45	0.45	0.40	0.40	0.30	0.30
1981	0.70	0.64	0.60	0.54	0.50	0.45
1982	0.81	0.67	0.69	0.57	0.58	0.45
1983	0.81	0.59	0.69	0.51	0.58	0.42
1984	1.00	0.66	0.84	0.55	0.70	0.46
1985	1.11	0.63	0.94	0.54	0.77	0.44
1986	1.11	0.56	0.94	0.47	0.77	0.38
1987	1.11	0.44	0.94	0.37	0.77	0.30
1988	1.11	0.33	0.94	0.28	0.77	0.23
1989	2.17	0.56	1.95	0.51	1.74	0.45

Source: Mtawali, K. M., *An Analysis of Characteristics of Households Facing Food Security in Malawi* (East Lansing, Mich., U.S.A.: Kellogg International System, 1989). Malawi, *Monthly Statistical Bulletin*, July 1991, Zomba, Malawi.

[a] Other areas include district headquarters and rural areas.

[b] Real wage rates are deflated by the Composite Retail Price Index.

and investment strategies increased the economy's vulnerability to external shocks. The heavy taxation on smallholder output reduced incentives for growth and had distortional effects on the economy. Nonpolicy or exogenous factors that contributed to the downturn include the falling of commodity prices, the oil price shock, trade route redirection, and drought.

The export-led growth of Malawi was largely dependant on three crops—tobacco, tea, and sugar. The commodity concentration ratios for these crops increased from 69 percent in the early 1970s to 83 percent in 1979. The high dependence on a few crops makes Malawi very vulnerable to adverse changes in weather and to international market conditions. The situation was compounded because Malawi, unlike her neighbors Zambia and Zimbabwe, does not have sizable extractive and manufacturing sectors. The fall in international prices for most commodities, especially tobacco, was a shock to the economy. The terms of trade, which had been positive for the early 1970s, declined by an average of 15.5 percent per year between 1977 and 1980.

The fall in terms of trade was compounded by the increase in the cost of imports and the redirection of trade routes to what has come to be known as the Northern Corridor. The war in Mozambique necessitated the closure of the traditional trade routes through the ports of Beira and Nacala. This meant increased costs of freight and insurance, and it is estimated that the economic cost of diversion stood at US$30 million in 1983.

Government investment, which had contributed significantly to the high growth rates of the 1970s, was not without costs. Between 1967 and 1979 the share of private consumption in total GNP declined by 23 percent. The concentration of investable resources on productive activities meant that public services were largely unattended (with the exception of defense). The allocation of public expenditure to health and community services was notably low in the 1970s; 2-3 percent for community services and 6-8 percent for health. On the other hand, allocation to economic services (material and service production activities) varied between 50 percent and 60 percent. The concentration on productive activities at the expense of social services was itself a cause of subsequent structural disequilibria.

Probably the most potentially explosive structural weakness as far as investment in 1970s was concerned relates to the question of sustainability. First, most of the investment went into low-return projects. Second, most of the investable resources were from external sources. Domestic savings were inadequate to cover the national investment requirements. Both public and private investments as well as the fiscal budgets were therefore covered by external borrowing that later exacerbated the debt problems of the country.

The problem of overdependence on foreign resources to finance domestic investments was complicated by the change in the nature and composition of foreign funding. From the initial postindependence period until the early 1970s, most of the external borrowing was at concessional rates. As time went on, however, borrowing from commercial markets became increasingly important. Interest rates on these commercial loans have been high (averaging 15 percent) and the repayment period short. Almost 20 percent of the budget went to interest repayment in the 1980s. Other factors that contributed to the structural disequilibria include the oil price shock of 1979, which resulted in an increased import bill for this and related commodities. The drought that hit Sub-Saharan Africa in 1980/81 adversely affected output and, hence, potential revenue and also necessitated a large importation of food, thus straining further the already undermined foreign-exchange situation. Movements in the balance of payments between 1981 and 1990 are shown in Table 7.2.

MAIN REFORMS INSTITUTED

The period since 1980 has seen Malawi undergoing or adopting several important reforms. The objectives of these reforms have been to reestablish stability in the balance of payments and regenerate the earlier growth momentum of the 1970s. The implementation of the reform has largely been financed by the International Monetary Fund (IMF) and the World Bank, with the IMF focusing on its traditional areas of assistance in exchange-rate and demand management, while World Bank assistance took the form of structural-adjustment loans (SALs). The reform processes over the decade have emphasized improved producer price incentives and removal of fertilizer subsidies; liberalization of markets, especially grain markets; and emphasis on adjustment of the exchange rate and rationalization of monetary and credit policies.

Price Incentives and Subsidy Removal

Pricing policy pursued in the smallholder subsector before 1980 was a significant element of implicit taxation of smallholder output. The consequences of such price distortions were manifest in the declining production of such major crops as cotton, rice, and groundnuts. Even

Table 7.2—The balance-of-payment problems, 1981-90

Balance-of-Payment Item	1981	1982	1983	1984	1985	1986	1987	1988	1989	1990
					(K million)					
Current account										
Merchandise trade (f.o.b.)	6.4	30.9	4.6	44.1	134.2	164.4	222.7	103.6	-93.1	-60.3
Exports	257.5	256.4	249.5	325.0	429.7	451.6	615.1	751.7	741.0	886.8
Imports	251.1	225.5	244.9	280.9	295.5	287.2	392.4	648.1	834.1	946.9
Nonfactor service	-96.7	-104.1	-140.8	-171.0	-227.4	-228.7	-284.4	-429.5	-546.4	-629.8
Receipts	36.8	38.6	44.5	48.4	45.3	39.7	50.0	72.6	76.1	84.0
Payments	133.5	142.7	185.3	219.4	272.7	268.4	334.3	502.1	622.5	713.8
Factor services (net)	-47.5	-31.5	-30.7	-50.4	-90.9	-122.0	-125.7	-137.9	-155.4	-141.4
Private transfers (net)	-1.3	-9.2	-11.6	-14.4	17.9	22.7	54.2	182.3	251.4	272.0
Current account balance	-139.1	-113.9	-178.5	-191.7	-166.2	-163.6	-133.1	-281.5	-543.5	-559.5
Capital account										
Long-term capital (net)	89.2	41.3	38.3	98.7	73.2	155.5	218.7	373.7	359.8	550.8
Government transfers	41.4	38.3	34.7	44.5	42.1	51.0	64.4	205.9	151.7	234.3
Government loans	31.9	23.7	28.4	87.8	47.8	114.7	139.9	132.5	158.7	236.7
Public enterprises	2.9	-17.5	-19.3	-23.5	26.5	15.0	10.3	30.1	38.5	65.8
Private enterprises	13.0	-3.2	-5.5	-50.0	9.8	4.8	2.1	7.2	10.9	14.0
Short-term capital plus errors and omissions	15.0	6.2	16.7	53.8	-23.4	-109.3	-50.0	42.8	69.6	8.7
Overall balance before debt relief	-34.9	-66.4	-123.5	-34.9	-116.4	-117.4	35.6	137.0	-114.1	0.0
Debt Relief	0.0	19.7	65.5	34.9	11.7	5.3	49.9	121.0	53.8	0.0
Balance after debt relief (equals change in net foreign assets of the banking system)	-34.9	-46.7	54.0	0.0	-104.7	-112.1	85.5	258.0	-60.3	0.0

Source: Malawi, Office of the President and Cabinet, *Economic Report* (Zomba, Malawi: Government Printer, various years).

maize output showed some erratic behavior. The fall in smallholder output in the 1980s reflects the policy inadequacies of the World Bank's Integrated Rural Development Projects approach. The assumption for these projects was that if infrastructure was in place, output would automatically respond. The relegation of pricing to secondary importance explains why the expected output levels were never realized. It is not surprising, therefore, that the initial policy reform in 1980 under SAL 1 emphasized the improvement of price incentives. Apart from increasing prices in general, price reform also aimed at raising the relative prices of export crops.

Although there was a policy attempt to drive prices toward export (import) parity levels, the manner in which these reviews were carried out created some problems in terms of inducing increased output and hence increased incomes, placing a burden on the treasury. During the early 1980s, prices were reviewed not on the basis of set objectives but rather as a reaction to some prevailing circumstance. The price reviews of 1981/82 and 1986/87 were reactions to declines in production due to bad weather and stagnant prices. The subsequent increases in production that followed these price increases did not reflect productivity improvements but rather a substitution of land for maize at the expense of other crops.

Another area of price reform concerned the removal of fertilizer subsidies. This issue became particularly prominent under SAL 2 when a time frame was actually set for the phased removal of such subsidies. The subsidy level, which stood at 22.6 percent in 1985/86, was to be completely removed by 1988/89.[3]

The conviction that production would not be affected by the resultant increase in fertilizer prices was based on the assumption that demand for fertilizer was highly inelastic so that any change in price would not significantly affect demand for the product. The subsidy-removal program was calculated based on the Beira and Nacala routes. Hence in terms of fertilizer-pricing through these routes, Malawi had virtually removed the subsidies on fertilizers by 1984. Subsidy elements since then reflect increased transportation costs resulting from trade-route diversion away from Mozambique.

The uptake of smallholder fertilizers increased during the 1980s despite reduced subsidies. The complicating factor when analyzing price response is the leakage of smallholder fertilizers to estates. The

[3] To cut down on import costs, high-analysis fertilizers were to subsequently replace the traditional types.

leakage estimates vary from 20 percent by the Ministry of Agriculture to between 35 and 50 percent by independent sources (Nathan Associates 1987). The major cause for the leakage of fertilizers to estates is the price differential for the product between the two subsectors because of subsidies on smallholder fertilizers. Another related reason is the convenience of the Agricultural Development and Marketing Corporation's markets and selling points relative to the outlets for estate fertilizers, for example, Optichem and Agricultural Trading Company.

DOMESTIC MARKET LIBERALIZATION

Another important reform related to the liberalization of the domestic market—especially for grain. Under the Agriculture (general purposes) Act of 1987, private traders were for the first time legally permitted to compete with ADMARC in the marketing of a range of smallholder agricultural produce. The main objectives of domestic market liberalization were to improve ADMARC's financial performance through its withdrawal from unprofitable markets and to improve market efficiency through increased competition.

The only requirements for participating as a private trader are an application fee of K 10 and a license fee of K 60. Private traders are allowed to operate only in designated markets. The choice of markets from the designated pool is up to the prospective private trader. The result has been that most private traders have concentrated their operations in the most convenient and profitable markets. In Lilongwe Agricultural Development Division, for example, almost all private traders obtained licenses to operate in markets within the Thiwi/Lifidzi rural development project in the 1988/89 marketing season. In the majority of cases, the areas deprived of private traders have been those affected by the closure of ADMARC markets and selling points (Mkwezalamba 1989). Thus there has been increased competition and efficiency in some markets but reduced market access for the peripheral areas.

Although conditions of entry are fairly easy, the growth in private trader participation has not been impressive (Table 7.3). A partial explanation for the slow growth of private trader participation may be that many people are already engaged in small-scale transactions in maize without a license. Also, there could be other potential traders who have so far been unable to participate because of lack of financing. Most of the existing private traders are actually already well established businessmen with their own trucks.

Table 7.3—Number of private traders registered by Agricultural Development Division, 1988-91

Agricultural Development Division	1988	1989	1990	1991
Karonga	...	7	1	1
Mzuzu	22	35	16	13
Kasungu	10	27	10	6
Salima	20	28	16	15
Lilongwe	113	128	90	169
Liwonde	109	224	145	165
Blantyre	99	417	241	199
Ngabu	14	51	24	42

Source: Internal documents of the Ministry of Agriculture, Lilongwe.

As far as pricing is concerned, the government still sets a floor price below which private traders are theoretically not allowed to buy produce. The private trader is, however, free to pay the producer a price higher than the floor price. In reality, it has been found that in some remote areas, traders have tended to operate on a cost-sharing basis with producers. Large trucks are hired and sent to areas with grain to be removed, and the farmers have agreed to bear some of the cost of the truck-hire rates (Kaluwa 1990). This effectively means that farmers receive a price lower than the ADMARC floor price. The government also sets a maximum consumer price that the private trader can charge. The maximum consumer price set for private traders seems to hold only during periods of normal supplies, and even then, consumers in deficit areas may face prices higher than the set maximum price. Private traders have sold maize at prices higher than K 45 per bag (1 bag equals 90 kilograms).

Preliminary information suggests that the most profitable option for the private traders is to sell the grain to the consumer. This option is difficult to them at present because they lack adequate facilities. Hence they usually dispose of the grain to ADMARC or processors within the first few months of buying the crop.

Since it is known that in most cases private traders tend to pay a higher producer price than ADMARC, the traders' margins based on a floor price might be slightly exaggerated. For example, in 1988/89 private traders paid a producer price ranging from K 16.37 to K 18.00 per bag between May and September, while the ADMARC producer price was K 14.85 per bag.

In terms of success of the reform, there is no doubt that most farmers have benefited from the choice of marketing channels and higher producer prices for their commodities. However, farmers in those areas where ADMARC has withdrawn and private traders have not moved in would appear to have been adversely affected in terms of both income and food security. Rather than blame the private traders, the government should develop infrastructure in such remote areas to encourage private traders to do business there.

The growth in privatization has been rather slow. Although the major bottleneck appears to be finance for private traders, the system of licensing, and allocating designated pools, also hinders participation. In spite of only a partial market liberalization, some of the benefits expected from the reform are being realized. The main problem area at present is the distribution of inputs to smallholders, an activity that the private traders have so far avoided.

Monetary and Fiscal Reforms

Monetary reforms emphasized the control and rationalization of credit interest rates, and currency adjustments to ensure competitiveness in the international markets. However, these reforms have been difficult to implement. The main problem arose from the government's need to finance its fiscal deficits, a need that drove it to increased borrowing on the domestic financial market. Government borrowing combined with borrowing by statutory agencies engineered the expansion in money supply. Between 1980 and 1988 money supply measured by cash and demand deposits increased at an average of 20 percent per year. Savings and time deposits also increased significantly.

The increase in money supply, coupled with devaluation of the Malawi kwacha during the 1980s, contributed to a worsening of inflationary conditions. Between 1976 and 1981 inflation averaged almost 10 percent annually, but between 1983 and 1988 the average annual rate of inflation stood at almost 17 percent. In 1987 inflation was as high as 25 percent.

The high levels of inflation combined with restricted access to imports and foreign exchange discouraged private-sector investment. This is evidenced by the decline in private-sector domestic assets from 59 percent of total assets in 1980 to 24 percent by 1987. The situation was arrested in 1988 when private-sector credit increased to 46 percent, while government's share declined from 60 percent to 37 percent. In short, for most of the 1980s, financial restraint stipulated in reform agreements was not adhered to by the government.

The main target of fiscal reform was the reduction and subsequent elimination of the deficit. This was to be achieved through policies that restrain public expenditure and the promotion of policies that enhance revenue generation. In an effort to increase revenue, a number of taxes, including charges on public utilities such as electricity and water, were raised. Restraining government expenditure proved to be a difficult exercise. Although government expenditure on the development account was reasonably well contained between 1980 and 1989, expenditure on current account kept increasing over the period (Table 7.4). The main reasons for failure to exercise restraint on the current account were the payments of interest charges and, in some cases, repayments of capital as well. Interest payments represented over 20 percent of total recurrent

Table 7.4—Gross fixed capital formation (GFCF) as a share of government expenditure, 1980-91

Year	Recurrent Expenditures	Development Expenditures	Total Expenditure	GFCF	GFCF as Share of Total Expenditure
		(K million)			(percent)
1980	207.64	174.90	382.54	248.7	65.01
1981	309.72	124.27	433.99	167.8	38.66
1982	278.45	139.58	418.03	181.7	43.46
1983	315.28	142.92	458.20	197.3	43.06
1984	416.79	138.38	555.17	222.7	40.11
1985	532.39	168.92	701.31	259.5	37.00
1986	547.03	186.13	733.16	252.2	34.39
1987	764.08	208.93	973.01	352.9	36.27
1988	847.54	342.07	1,189.61	524.0	44.05
1989	1,103.65	295.29	1,398.94	699.6	50.01
1990	1,168.74	408.25	1,576.99	620.0	39.31
1991	1,331.35	437.20	1,768.55	950.0	53.72

Source: Malawi, Office of the President and Cabinet, *Economic Report* (Zomba, Malawi: Government Printer, various years).

expenditure between 1981/82 and 1988/89. The failure to cut recurrent expenditure resulted in policy-induced inflation, which also affected private-sector investment incentives. Restraint on the development account affected the level of economic activity and the provision of social services such as health and education. It is not surprising that the initial gains from the reform process realized in 1983 and 1984, when growth rate recovered to 3.6 percent and 4.5 percent, respectively, were lost by 1986.

OTHER REFORMS

Under the Agricultural Sector Adjustment Credit, the government has legalized (on an experimental basis) the growing of burley tobacco by some smallholders. This crop had been a monopoly of estate producers. Also, the research for a flint hybrid maize with storage and processing qualities comparable to the local variety was emphasized, and such a variety has actually been developed and is undergoing multiplication.

Another notable reform feature was the inclusion of questions relating to nutrition and poverty in general in the years after 1985. While various studies have been undertaken on the subject, especially on nutrition problems, the donors and the government are trying to devise strategies so as to "adjust with a human face." The issue of very high population growth and strategies to reduce this growth are also assuming increasing importance in the country.

SOME GENERAL OBSERVATIONS

As stated ealier, one of the important elements of policy reform through the 1980s was directed at liberalization of various sectors in the economy. One such area is the marketing of smallholder agricultural produce. Although the participation of private traders is still relatively small and information on their activities is limited, the absence of private-trader participation in the sale of inputs needs to be investigated. Traders have not replaced ADMARC in the markets that were abandoned as a result of the rationalization of economic activities. Both ADMARC and private traders are competing in accessible markets. This has had adverse effects in most remote areas in terms of access to inputs and markets.

CONCLUSION

In agriculture, reforms have focused on price and market liberalization; removal of subsidies, especially on fertilizers; and slowdown in the expansion of the estate sector while trying to provide funding to improve the access of smallholders to medium- and long-term credit. Economywide policy reforms have emphasized reduced public spending and a reduction in the size of the civil service, exchange-rate adjustment to ensure competitiveness, restrained credit and rationalized interest rates, tax breaks and tax hikes where necessary, and efforts to ensure profitability of parastatal bodies.

Finally, some of these reforms have been carried out for more than 10 years now, but as noted above, the economy is still largely dependent on foreign financing. There is certainly a need to take stock of how effective these reforms have been and to modify them, depending on the degree of contribution of an individual reform policy to the adjustment process.

REFERENCES

Kaluwa, B. M. 1990. Private traders' response to food marketing liberalization, market integration and economic efficiency. Paper presented to a National Workshop on Food Security, Socioeconomic Status and Nutrition of the Rural Population in Malawi, December, Chancellor College, Zomba, Malawi.

Kydd, J., and R. Christiansen. 1987. *Malawi's agricultural export strategy*. U.S. Department of Agriculture Staff Report No. AGES 70224. Washington D.C.: USDA.

Malawi. 1991. *Monthly Statistical Bulletin* (July). Zomba, Malawi: Government Printer.

Malawi, Office of the President and Cabinet. Various years. *Economic Report*. Zomba, Malawi: Government Printer.

Mkandawire, R. 1990. *Beyond dualism: The changing face of the leasehold estate sub-sector of Malawi*. Lilongwe: USAID.

Mkwezalamba, M. M. 1989. The impact of the liberalization of smallholder agricultural produce pricing and marketing in Malawi. A Draft Final Report submitted to the Ministry of Agriculture, Lilongwe.

Mtawali, K. M. 1989. *An analysis of characteristics of households facing food security in Malawi*. East Lansing, Mich., U.S.A.: Kellogg International System.

Muir, K. 1982. *Agricultural marketing and price policy in Malawi*. Working Paper AEE 5/82. Harare: University of Zimbabwe, Department of Agricultural Economics and Extension.

Nathan Associates. 1987. The impact of fertilizer subsidy removal program on smallholder agriculture in Malawi. Report presented to the Ministry of Agriculture and USAID, Lilongwe.

Sahn, D. E., J. Arulpragasam, and L. Merid. 1990. *Policy reform and poverty in Malawi: A survey of a decade of experience in Malawi*. Cornell Food and Nutrition Policy Program, Monograph 7. Ithaca, N.Y., U.S.A.: Cornell University Press.

Part III
Effects of Domestic Policy Reforms on Food Security

8

Is Market Liberalization Compatible with Food Security? Storage, Trade, and Price Policies for Maize in Southern Africa

Thomas C. Pinckney

Liberalization of both internal and external markets is promoted widely by multilateral and bilateral donors as an important step in the revitalization of African economies. Indeed, over the last six years many African countries have taken significant steps toward freeing exchange controls, trade barriers, and internal restrictions on markets. An increasing number of African governments are recognizing the important role that market prices can play in providing the appropriate signals for investment and consumption decisions.

Governments moving toward a more liberalized economy have encountered a difficult dilemma in markets for staple foods. Liberalization of the staple-food market ideally should stimulate increased food production and investment in private marketing services; unfortunately, these benefits occur only (or are frequently perceived to occur only) in the long run. In the short run, liberalization often is thought to result in declines in real income, rapid fluctuations in price, and possible political chaos.

Some features of African food markets could make this fear a reality. African countries—especially Zambia and Zimbabwe—have exceptionally high levels of variability of staple-food production. The present drought is only the latest evidence of this instability. Production variability combined with the usual inelastic demand curve for the primary staple must lead to high levels of price fluctuation from year to year unless availability is affected by changes in stocks or foreign trade. But storage and trade are expensive propositions in the region. Storage of foodgrains requires both a high capital cost up front in order to build proper facilities and high costs each year to hold the stock (typically 15 to 25 percent annually of the value of the stock). Poor transportation networks and continued civil strife in Mozambique make the difference between import and export parity for countries in southern Africa close to US$200 per metric ton (Koester 1986). Thus, there is no obvious policy initiative to recommend to liberalizing countries concerned about price instability.

This paper, therefore, addresses the following question: Given large, random swings in domestic food production, inelastic demand curves, and the high costs of storage and trade, how can governments gain the efficiency advantages of moving toward freer markets and still protect producers and consumers from the large costs of fluctuating prices? Can economic modeling help in the design of policies that stabilize prices and incomes at a reasonable cost?[1]

The question is applied to Zambia, Zimbabwe, and Malawi as well as the three countries together as a region. Historically, Zimbabwe has been a maize exporter, Zambia an importer, and Malawi self-sufficient, although all three countries have both imported and exported in the last 10 years. These historical differences lead to a consideration of a wide range of policy issues, making generalizations of the results easier.

Other analysts have examined these issues in part. Studies concerning countries in the region include two papers by Buccola and Sukume (1988, 1991). Both papers apply the results of a theoretical analysis to Zimbabwe. National utility functions are maximized in a two-period model. Primary conclusions of importance to this study are (1) an "optimal" stock level can be chosen only in conjunction with price; and (2) in the mid-1980s, neither risk aversion nor utility maximization could explain the very high levels of stock held by Zimbabwe. The interdependence between stocks and price is elaborated upon in this study, while the second point—the size of the interannual supply stabilization stocks—will be studied from a different perspective and the results compared.

The Oxford Food Studies Group has developed a model of grain-market intervention and calibrated the model with numbers from Zimbabwe (van der Geest 1991). The model is considerably more complex than the one developed here, with a different set of advantages and disadvantages. Its main advantages are for analyzing within-year policy issues and attempting to model all of the important factors that

[1] Such policies have both interannual and seasonal components. The scope of this paper is limited to the interannual components. See Pinckney (1990) for an analysis of Malawi using a similar framework that includes seasonal issues. Duncan et al. (1990) examine seasonal issues for Zimbabwe using a framework that is similar in some ways. These two analyses relate national, seasonal price fluctuations to fluctuations in national supply that are beyond the control of the government. In some cases, government policies actually increase the seasonal price fluctuations faced by rural consumers; see Jayne et al. (1990) and Jayne and Chisvo (1991a) for analyses of how internal market reform in Zimbabwe could help to mitigate these policy-induced fluctuations.

have an impact on supply and demand for the staple food. It is not an optimization model and thus cannot be used to answer many of the questions about policy design asked here. Nevertheless, it would be useful to test the types of policies generated by the present model in the more detailed context of the Oxford Food Studies Group model.

This study grows out of a series of IFPRI studies (Pinckney 1988, 1989) on Kenya and Pakistan. The methodology uses a fairly simple model to allow for optimization in a stochastic framework. In this context, optimization is useful for screening out inferior types of policies, for indicating the elements of efficient policy design, and for accurately measuring trade-offs between objectives.

The rest of the paper is organized as follows: first, the price stability goal itself is examined as to whether or not it is a reasonable goal for government. Next, some general principles of intervention that hold for all countries are laid out. This is followed by a summary of the model. The question is then asked, Can the legitimate goals of government be achieved adequately in southern Africa by free markets? Since model results suggest that the answer is "no," further results are then reported, suggesting ways that governments can intervene to stabilize markets at low cost in the three countries of interest. Taking these formal model results, implementation issues are discussed for the countries in question. A final section presents the conclusions of the study.

SHOULD GOVERNMENTS STABILIZE PRICES FOR STAPLE FOODS?

Most officials in less-developed countries take the stabilization function of governments for granted. The question is not whether or not government should intervene in the market, but whether government should mandate a single price or just influence the price. But there is a long and distinguished debate among academic economists concerning the welfare effects of price stabilization. The final result of that work has been to show (1) that costless price stabilization may or may not be beneficial, depending on the shape of the demand curve; and (2) that the welfare costs of price instability, when they exist at all, are relatively unimportant.[2] This research would seem to imply that

[2] See Waugh 1944, Oi 1961, Massell 1969, Samuelson 1972, Newbery and Stiglitz 1981, Stiglitz 1987, and Wright and Williams 1988 for important contributions to the debate.

governments should never spend resources to stabilize prices.

This tradition of analysis, however, is completely static, concentrating on the measurement of "welfare triangles," the efficiency gains and losses that accrue at one point in time when prices vary to producers and consumers. But the costs of price instability are likely to be dynamic, not static. Timmer has argued that these dynamic costs include

> (1) displaced investments in physical capital at the farm level, the marketing sector, and the industrial sector; (2) substitution of consumption and leisure for savings and work; (3) biases in investments in human capital for the farm agent and intergenerationally in children; (4) the transactions costs consumers face in reallocating budgets when prices change; (5) the welfare gains from a psychic sense of food security . . .; and (6) the feedback from this sense of security to a stable political economy, which reinforces investors' willingness to undertake long-term (and hence risky) commitments [numbers added] (Timmer 1990, 25).[3]

Seen in this context, the desire of governments to stabilize prices is quite rational. Timmer's first three points all are consistent with the well-known negative impact of price instability on the output of risk-averse producers. Indeed, Bigman (1985) several years earlier had reported simulation results that suggest that the benefits of supply response to price stabilization far outweigh the "welfare triangle" effects. The fourth point is particularly important for staple foods, which may constitute 20-30 percent of a poor consumer's expenditure. A doubling of prices for the staple may require a reallocation of a quarter of total expenditure; the transactions costs involved are undoubtedly high.

Points (1), (2), (3), and (6) all are concerned with investment and growth. No static analysis can capture such effects. Thus, in trying to stabilize prices for staple foods, governments are in effect trading off some short-term allocative inefficiency against a reduction in short-run transaction costs and an increase in long-term growth. The difficulty of

---

[3] See Timmer 1990 for a detailed discussion of these costs. Also, see Dawe 1990 for a more theoretical assessment of the long-run impact of food price instability on the macroeconomy.

the trade-off is twofold: to ensure not only that the benefits exceed the costs, but also that allocative inefficiency is short term and does not bias investment in unprofitable ways in the long run.

Thus, governments do indeed have a legitimate interest in stabilizing prices, notwithstanding the long debate in the economics journals. The following section discusses some general principles of intervention in food markets that hold for all governments. It is possible, however, that in any particular market for staple foods, an unregulated market would stabilize prices adequately, so that no government intervention is required. Whether or not this holds for the three southern African countries under consideration is discussed later in the chapter.

SIMPLE ANALYTICS OF FOOD-MARKET INTERVENTION

When the production of a staple food fluctuates dramatically from year to year, consumption will fluctuate also unless supplies are removed from the market in surplus periods and brought into the market in deficit periods. This simple statement can be formalized through the use of a set of identities. For a particular year in any country, total supply must equal total demand. There are three possible sources of supply—production, opening stocks, and imports—and three possible sources of demand—consumption, closing stocks, and exports.[4] In equation form, this can be stated

$$Q_t + S_{t-1} + M_t = C_t + S_t + X_t, \qquad (1)$$

where the subscript refers to the year, and

Q = production,
S = closing stocks,
M = imports,
C = consumption, and
X = exports.

[4] In this statement and the following equations, production is assumed to be net of losses, and food aid is included with imports.

Solving for Q yields

$$Q_t = (S_t - S_{t-1}) + (X_t - M_t) + C_t. \quad (2)$$

That is, production equals the change in stocks plus net exports plus consumption.

Equation (2) implies that when production, Q_t, fluctuates from one year to the next, at least one of the three terms on the right-hand side will also have to fluctuate. In other words, production variability must be translated into stock variability, trade variability, or consumption variability.

A country that is an exporter in a normal production year can buffer production fluctuations by reducing exports, provided the production shortfall is less than its usual exportable surplus. Similarly, a normal-year importing country can increase imports in a bad year. The production shortfall could translate into tightness in the supply of foreign exchange—and if domestic prices are not at parity with world prices, or if the exchange rate is misaligned, the country may want to adjust domestic prices in order to decrease consumption during the shortfall. In the case of neither the importer nor the exporter, however, is it likely that stock changes will play any role in buffering such production fluctuations, assuming that the country is small in world markets. The exporter would have to forego exports in year t − 1 if it is to hold stocks for a possible shortfall in supply in year t. Thus, although holding stocks may allow the country to maintain exports in the face of the production shortfall in year t, the net effect is to move exports from year t − 1 to year t, while incurring a year's worth of storage charges. Such an operation would lose money unless there is a very large movement in world prices between the years. A similar analysis would hold true for the importing country.

The situation is considerably more complicated when a country is self-sufficient in a normal production year, or is a normal-year exporter (importer) that occasionally imports (exports). This results primarily from the costs of engaging in foreign trade. If a country could buy maize in a deficit year at the "world price" and sell in a surplus year at the same price, the situation would be the same as for the normal-year exporter; production fluctuations could be translated into trade fluctuations at little cost. But because of transport and handling costs, the price received domestically for exported grain is considerably less than the world price, while the price paid domestically for imported grain is considerably more than the world price. With a difference between export and import parity of approximately $200 per ton in southern Africa, if a country in the region has a surplus in year t, and knows that it will be in deficit in year t + 1, it is profitable to store the commodity for one year, since storage costs are only about US$25 per ton.

The problem is that next year's crop is not known, and if the stock has to be stored for more than a couple of years, the government is losing more money than if it were to depend on imports. Thus, there is no obvious answer to the question of whether or not stocks for interannual supply stabilization should be held by a normally self-sufficient country.

The simple identity in equation (2), however, does provide at least four clues as to the general type of policy that will be most effective. First, a "food security reserve" that ends every market year at the same level is totally ineffective at buffering production fluctuations. In equation (2), it is the change in stocks that counteracts the shortfall in production. Thus, if stocks do not change, whether they begin and end the year at 0 or 3 million tons, all of the production decline will be translated into increased imports or decreased consumption. Such a reserve may serve a different purpose; countries that do not normally import may require a buffer to ensure that government stocks do not go to zero after imports are ordered but before they arrive in the country.[5] But this type of reserve does nothing to stabilize consumption across years. A true interannual supply stabilization stock will change in size from year to year, with the change serving to offset production fluctuations. The year after a bumper harvest, this type of stock will either be the same size or larger; the year after a drought, this type of stock in many cases should disappear.

Second, when analyzing interannual stocks, the correct question to ask is "Given the present and expected future values for production, stock level, and world price, how much of the staple should be consumed, exported, and carried out?" This question is markedly different from the alternative, "How much stock should the government hold at the beginning of every market year?" The first rather than the second is the question facing governments, since it includes the

---

[5] Such stocks have been termed "import buffer stocks" in Pinckney (1989) and have been analyzed by Duncan et al. (1990) for Zimbabwe. They should have large seasonal components, moving to their lowest level at harvest time and to their highest level three to five months prior to harvest. Analyzing the appropriate size of such stocks requires consideration of a very different set of issues from analyzing the size of interannual supply stabilization stocks.

economics of possible present and future uses of the stock. The first question clearly will lead to different answers in different years. If export prices are low and the country is in surplus, the optimal closing stock may be high; if the country is importing in the present year, the optimal level for interannual supply stabilization stocks will be zero.

Third, domestic price policy is intimately linked to storage issues. In equation (2), if domestic prices are allowed to change in response to fluctuations in production, consumption will change, leading to less of a need for stock or trade variability. The extent to which domestic consumption varies with domestic production is thus important in the formulation of a storage strategy.[6]

Fourth, international supply stabilization stocks and foreign trade in the staple are substitutes for each other. Production fluctuations can be buffered by changes in storage or by international trade. If international markets can be made reliable, and if—as is frequently the case—relying on trade is cheaper than relying on stocks, there is no reason to hold interannual supply stabilization stocks.[7]

These four clues, then, give some guidance as to the type of intervention governments should pursue in food markets if such intervention is warranted by excessive price instability. There are two decisions facing the intervening government that are best analyzed sequentially. First, for any given production deviation from normal, how much should domestic prices adjust to a production deviation, thereby allowing consumption to vary with production? Second, of the remaining production variability, how much should be assigned to stock variability, and how much to trade variability? The analysis above suggests that the answers to these questions are likely to depend on world prices and opening stocks. The next section of the chapter formalizes these ideas in a simple model. Whether or not governments

---

[6] This statement refers only to interannual stocks, but the issue is equally important in an analysis of seasonal stocks. The larger the expected seasonal price increase, the more seasonal stocks will be held by farmers and traders. Government policy that dampens or removes seasonal price rises can be both expensive for the government and inefficient for the economy, as the government will be forced to transport and store much of the large amount of the staple that must be held after harvest and, in the worst case, transport and sell it back to the very farmers that grew the crop. Zimbabwe's price policies have encouraged such behavior (Jayne and Chisvo 1991b).

[7] Governments may still have a rationale for holding the other two types of stocks, import buffer stocks and seasonal stocks, which may require some holding of stocks across market years. The purpose of such stocks, however, is not to buffer changes in production.

should intervene at all is then analyzed by using the model to estimate the level of price instability that arises from free markets.

AN INTERANNUAL SUPPLY STABILIZATION MODEL

This model attempts to simplify the problem confronting food policymakers in order to shed light on the important elements of supply stabilization. The simplifications are described here.

Each year t, a quantity of maize is produced, Q_t. This quantity is random, and in year t - 1 no one can predict the value of Q_t. In year t, however, Q_t is known. The government enters year t with S_{t-1} in storage, with S_{t-1} defined as the closing stocks of year t - 1 (as above). Also, the world price, WP_t, is known in year t.

These three variables—production, stocks, and world price—are the only parameters that change from year to year in the model. All other parameters—including the costs of importing and exporting any particular amount of maize, the demand elasticity of maize, the cost of storage, and the standard deviation of production—are constant throughout all years.

Once the values of the three variables are known, the government has two policy decisions to make. First, it must determine how much to buy or sell domestically. Since each year is a single point in time in the model, what matters is net government purchases or sales during the entire year. Thus the government may be removing maize from the market in surplus years and adding to supplies in deficit years. The variable for net purchases by the government is termed NP_t.

The second government decision is how much to import (M_t) or export (X_t). In a year of a production shortfall, the foreign trade decision may be determined by the domestic market decision; if net sales domestically are greater than opening stocks, imports must be large enough to make up the difference. In surplus years, however, the question is how much of the available supply to export, and how much to keep in storage in case next year's production is low. Note that closing stocks S_t are determined by opening stocks S_{t-1} plus net purchases NP_t plus net imports, and thus do not constitute a separate decision variable. The government has control over closing stocks, but has only two rather than three variables that it can control.

The government's decision problem, then, is to determine the "best" amount of domestic net purchases, exports, and imports given any possible combination of domestic production, world price, and opening stocks. But what criteria should be used in order to decide which policy is best?

The standard objective function in analyses of interannual supply stabilization has been the maximization of present and expected future values of producer/consumer surplus.[8] It is well known that an optimization model with such an objective function describes the free market solution (Gustafson 1958). This characteristic is used in the next section to estimate the level of price instability for southern African countries in the absence of government intervention.

This type of objective function has been used to measure the benefits or costs of price stabilization in the classic literature on the subject discussed above, and, as mentioned there, does not capture the dynamic costs of price instability. One approach to considering these costs in the objective function would be to measure the costs directly, and to subtract these costs from consumer/producer surplus. Measuring these costs, however, is a particularly complex task, with each step along the way highly debatable. Instead, the objective function proxies such costs by subtracting the product of a weight and the squared deviation of price from its target. The deviations are squared, since it is likely that the dynamic costs will increase at an increasing rate as prices become more unstable. They are weighted, since the relationship between these costs and producer/consumer surplus is unknown; by varying the weight, it is possible to map out the possible combinations of cost and price stability and to examine how the optimal policy changes as price stability becomes more highly desirable. This gives clues to the design of storage and price policies. In addition, there are some characteristics of all the optimal policies, regardless of the weight on price stability, that differ from existing methods of intervention.

---

[8] Note that in this model where planned production does not vary from year to year, production costs are constant, thereby making changes in producer/consumer surplus equal to change in consumer surplus alone. In order to estimate the importance of this simplification, a version of the model was tested in which production was allowed to respond to changes in expected price with a supply elasticity of 0.3 in the short run and 0.6 in the long run. The added complexity had virtually no impact on model results in all areas except one: stock levels. Average stock levels with supply response were about two-thirds of those without supply response. Supply response makes such little difference because expected prices are much less variable in these markets than actual prices, with the ratio of the coefficient of variation for expected prices to the c.v. for actual price ranging from 4 percent (for Malawi) to 25 percent (for Zambia). Most of the price variability results from production fluctuations and world price movements that are unknowable at planting time. Since supply response is generally unimportant, all results reported below are for the simpler model in which planned production is constant. Thus, estimates of average stock levels should be considered to be on the high side.

Note that in this annual model, the price stability goal is measuring fluctuations in the annual price around a target. The price must be measured at one point in the marketing chain; it makes most sense to think of this price as the average price paid to farmers at harvest time. Average consumer prices for the year can be at a different level but are assumed to move proportionately in the same direction as the farmgate price. This model specifically is not concerned with seasonal price variations, since each year is considered a single point in time.

The individual functions are as follows. Production is an exogenous, stochastic, normally distributed random variable. Thus, the costs of government policy are modeled for good, bad, and normal production years. There is no supply response to price in the model.[9]

The demand curve is assumed to have constant elasticity. Food aid, which covers a set proportion of the shortfall in domestic supply in bad years, is received by the government in the year of the shortfall. Food aid quantities turn out to be relatively small in the model on average.

The existence of a regional market paid for primarily by food aid agencies is modeled by allowing a set amount of exports in any one year at a markedly higher price. This turns out to be an important assumption of the model, as discussed below. Additional exports to the world market fetch a much lower price.

The scarcity of foreign exchange is modeled by valuing proceeds in foreign currency—such as exports—as 10 percent more valuable than proceeds in domestic currency.

Possible constraints on transporting imports are modeled by increasing the transportation charges if imports plus food aid exceed a specified amount. Costs for imports are linear until they reach this amount; at higher levels of imports, transportation costs increase exponentially.

World price is modeled as a random walk, with movements independent of domestic production. Thus, it is not possible to speculate successfully on world price changes.

The model is optimized through the standard stochastic dynamic programming technique of backward recursion. This produces an optimal "policy," where the policy is a prescription of the most appropriate domestic price and amount of foreign trade for every possible combination of production, world price, and opening stocks considered in the model. The reported results (shown in following sections) are outcomes of simulating a linearly interpolated version of

[9] See footnote 8 for a discussion of a test model that included supply response.

the optimal policy over 500 10-year cycles.

This concludes the brief model description.[10] Values of key parameters for Malawi, Zambia, Zimbabwe, and the region are presented in Table 8.1[11]

Table 8.1—Main assumptions of the model

Assumption	Malawi	Zambia	Zimbabwe	Region
Coefficient of variation of production (percent)	7	20	30	19
Demand elasticity for maize	-0.3	-0.3	-0.3	-0.3
Food-aid trigger (1,000 metric tons)[a]	1,130	1,020	1,450	3,490
Costs of exporting maize per metric ton (U.S. dollars)	80	80	80	80
Premium paid on regional exports per metric ton (U.S. dollars)	120	120	120	120
Maximum regional exports (1,000 metric tons)	60	100	120	200
Costs of importing maize per metric ton (U.S. dollars)	140	100	80	100
Maximum imports for linear import cost (1,000 metric tons)[b]	200	600	1,000	1,000

[a] One-half of the difference between the food-aid trigger number and the total of production and stocks is received in the year of the shortfall at no cost to the country.
[b] When imports plus food aid are greater than this amount, the cost of importing increases exponentially.

---

[10] A fuller description of a similar model can be found in Pinckney (1988, 1989). The main difference between those models and this one is in the objective function, where the earlier publications minimized a government cost function that included a price stabilization goal rather than a welfare function. The main advantage of the earlier method is in calculating directly the trade-off between government expenditure and price stability, a major goal of the earlier work.

[11] The parameters were chosen after examining available data, at least through 1989, for each country. Some parameters, such as the demand elasticity for maize, were chosen at a reasonable value rather than being based on existing estimates, because the estimates were deemed to be quite poor. Some sensitivity analyses to the values of key parameters have been conducted but are not reported here. Major conclusions do not appear to be sensitive to changes of 20 percent in the values of these parameters.

DO FREE MARKETS PRODUCE STABLE PRICES?

The model described above can be used to estimate the results of freeing all markets for maize in the countries under consideration by setting the weight on price stability to zero. Four models were run, one each for Zambia, Zimbabwe, and Malawi, and one for the three countries together. The aggregate model assumes that there is perfectly free trade among those three countries. Results are shown in Table 8.2.

All model solutions show that prices would be quite volatile in the absence of government intervention, with the coefficient of variation of price ranging from 22 percent for Malawi to 38 percent for Zambia and 51 percent for Zimbabwe. For the free regional market, price variability would remain high at 44 percent. Thus, for the region as a whole, a price deviation from the target of 44 percent or more would be expected in one out of every three years, and a deviation as large as 72 percent would be expected in one out of every 10 years. The level of price instability for Zambia, Zimbabwe, and the region would unquestionably have many of the negative dynamic consequences discussed above. Malawi's case is less clear, as in 9 out of 10 years the price would be within 36 percent of the target. The benefits of increasing price stability will clearly offset the cost of stabilization only if these costs are small. This is examined below.

Despite this high level of price instability, average interannual supply stabilization stocks are quite low, with virtually none held in Malawi and less than 100,000 tons held even for the region as a whole. Average stocks are somewhat misleading, however, since for many years stocks are zero. Zimbabwe, with the highest average stock level for the individual countries, is an interesting example. Although stocks average 73,000 tons, the median stock is zero. Stocks are less than 10,000 tons in 54 percent of all years, and less than 100,000 tons in 73 percent of all years. For only 9 percent of years are stocks greater than 300,000 tons, and they are less than 700,000 tons in virtually all years.

Note that these stocks are held by private agents when price expectations suggest that they will profit from holding stocks. Government is not intervening in this market at all. Stocks are held in the model when production is high, world prices are low, there is no further opportunity to export at high prices to the regional market, and domestic prices have fallen to low levels.

Instability of farm income from maize—here calculated as total production times the price—is also high under free markets, although lower than price instability, since some of the price deviations offset production fluctuations. The coefficient of variation for farm income

Table 8.2—Impact of free markets

Country	Coefficient of Variation: Price	Coefficient of Variation: Farm Income	Stocks	Commercial Imports	Food Aid	Exports	Foreign-Exchange Earnings	Profits of the Marketing Sector
	(percent)		(1,000 metric tons)				(US$ million)	
Malawi	21.6	15.2	5	0	0	59	13.4	4.4
Zambia	38.1	21.4	26	6	13	76	16.2	8.3
Zimbabwe	50.9	28.5	73	27	34	124	18.8	14.0
Region	44.0	25.5	91	16	26	202	32.9	11.5

Note: All values other than coefficients of variation are average annual values from the simulation.

from maize ranges from 15 percent for Malawi to 21 percent for Zambia, 26 percent for the region, and 29 percent for Zimbabwe.

The regional market created by food-aid donors is an important determinant of these results. Indeed, all of the countries including Zambia, a traditional importer, become normal-year exporters under free markets and the opportunity for food-aid exports. Average net exports across years are close to 60,000 tons for each of the three countries, as free markets lead to higher average prices, particularly in Zambia and Malawi, in order to take advantage of profitable exports. Consequently, each of the three countries has net earnings of foreign exchange on the maize account of between US$13 million and US$19 million. Profits in the marketing system as a whole are less once domestic costs of storage and net purchases of maize are included; these profits range from US$3.1 million for Malawi to US$12.2 million for Zimbabwe.

Thus, free markets yield profits for maize marketing and earn foreign exchange for the country at the cost of highly unstable maize prices. Although some maize is stored across years, significant amounts are stored only rarely, even with these high levels of price instability, since it is impossible to predict most movements in the domestic price.[12] Clearly, a reduction in price and income instability would be desirable if it could be accomplished at low cost. The next section examines the types of policies that can reduce price instability, and the costs of doing so.

LOW-COST POLICIES FOR STABILIZING PRICES: WHAT DO OPTIMIZING MODELS SUGGEST?

Basic Results

The results of gradually increasing the size of the weight on price stability in the objective function of the optimizing models for Malawi, Zambia, Zimbabwe, and the region are presented in Table 8.3. To put the coefficient of variation for maize prices into historical context, note that the coefficient of variation for real, detrended producer prices in Malawi from 1980 to 1988 is about 11 percent; for Zimbabwe from

---

[12] See footnote 8 for a discussion of expected prices.

Table 8.3—Impact of price stability on welfare, cost, stocks, and trade

Country	Coefficient of Variation: Price	Coefficient of Variation: Farm Income	Reduction in Welfare	Stocks	Commercial Imports	Food Aid	Exports	Foreign-Exchange Earnings	Profits of the Marketing Sector
	(percent)		(US$ million)	(1,000 metric tons)				(US$ million)	
Malawi	21.6	15.2	0.0	5	0	0	59	13.4	4.4
	20.1	13.8	-0.1	7	0	0	59	13.2	4.4
	18.3	12.1	0.0	10	0	0	59	13.1	4.3
	16.2	10.1	0.2	16	1	0	58	12.9	4.0
	14.0	8.4	0.2	22	1	0	58	12.6	3.7
	11.2	6.5	0.8	28	2	0	61	12.8	3.1
	9.1	5.1	0.3	37	2	0	60	12.4	2.6
	6.1	4.3	1.0	44	4	0	62	12.3	1.8
	3.8	4.9	0.7	49	5	0	63	11.9	1.1
Zambia	38.1	21.4	0.0	26	6	13	76	16.2	8.3
	35.8	19.4	-0.2	32	9	12	76	15.5	7.7
	32.7	17.7	-0.2	36	12	12	74	14.6	7.1
	28.4	14.8	0.3	46	15	12	74	13.8	6.2
	23.8	12.6	0.7	60	18	11	74	13.1	4.9
	19.9	11.7	0.8	73	21	10	75	12.4	3.6
	16.1	11.8	1.1	86	24	10	76	11.9	2.3
	14.1	12.1	0.8	94	25	10	78	11.9	1.6
	10.5	13.5	1.5	107	27	9	80	11.6	0.1

(continued)

Table 8.3—Continued

Country	Coefficient of Variation: Price	Coefficient of Variation: Farm Income	Reduction in Welfare	Stocks	Commercial Imports	Food Aid	Exports	Foreign-Exchange Earnings	Profits of the Marketing Sector
	(percent)		(US$ million)	(1,000 metric tons)				(US$ million)	
Zambia	7.2	15.6	1.7	117	29	9	84	11.6	-1.4
(continued)	4.3	17.2	1.3	124	31	9	86	11.2	-2.7
Zimbabwe	50.9	28.5	0	73	27	34	124	18.8	14.0
	44.4	25.7	1.0	87	36	33	123	16.2	10.8
	40.3	23.7	1.0	95	43	33	124	14.5	8.5
	35.7	21.7	1.7	117	48	31	125	13.0	5.8
	31.1	20.4	2.5	135	54	30	130	11.8	2.6
	26.9	20.1	2.8	150	60	30	136	10.6	-0.6
	22.4	20.4	3.2	170	64	29	143	9.9	-3.9
	18.5	20.8	3.0	190	68	28	150	9.7	-6.6
	14.3	22.4	3.8	196	70	27	162	9.7	-9.8
	11.2	23.5	2.4	214	73	26	164	8.9	-12.2
	6.5	26.2	4.7	232	78	25	171	7.7	-16.6
Region	44.0	25.5	0	91	16	26	202	32.9	11.5
	39.7	22.6	4.4	116	26	25	200	25.5	4.1
	36.1	20.0	6.2	137	37	24	203	17.7	-4.4
	32.4	17.4	5.3	170	46	23	203	11.3	-11.7
	29.0	15.3	5.1	208	52	22	206	6.8	-17.9

(continued)

Table 8.3—Continued

Country	Coefficient of Variation: Price	Coefficient of Variation: Farm Income	Reduction in Welfare	Stocks	Commercial Imports	Food Aid	Exports	Foreign-Exchange Earnings	Profits of the Marketing Sector
	(percent)		(US$ million)	(1,000 metric tons)				(US$ million)	
Region	26.4	13.8	6.0	238	58	21	211	2.3	-24.5
(continued)	22.6	12.4	8.7	274	66	21	223	-2.6	-33.4
	17.0	11.5	10.3	348	74	19	247	-4.1	-42.5
	12.0	12.5	9.1	399	81	18	271	-4.1	-49.9
	8.2	14.4	7.0	451	88	17	288	-4.1	-55.7
	5.9	15.8	9.1	463	93	16	299	-8.4	-64.1

Notes: All values other than coefficients of variation are average annual values from the simulation. The "reduction in welfare" is the reduction in consumer surplus less marketing costs of the policy in the present row compared with the policy in the previous row. The first row for each country repeats the free-market result from Table 8.2.

1981 to 1990 it is about 8 percent.[13] Several common patterns can be discerned in the table.

The initial reductions in price variability are relatively low-cost; indeed, for Zambia and Malawi, price variability can be reduced considerably at virtually no welfare cost. In each case, there is a significant increase in the costs of marketing, but these costs are offset to a large extent by higher consumer surplus. This higher surplus results primarily from the price stabilization reducing prices in deficit years more than it raises prices in surplus years, thus lowering average prices somewhat. Thus, it would appear that even for Malawi some reduction in price variability would be welfare-enhancing, although the exact benefits are unknown, since the costs of the first steps in the reduction are quite small. More than half of the increase in marketing costs results from increased losses (or decreased profits) on foreign trade. Average exports, however, either change very little or actually increase, highlighting once again the dominance of the profitable export market over other factors. The change in costs arises from a combination of three factors: first, a reduction in food aid, since average stocks are higher; second, an increase in commercial imports to limit price rises in deficit years; and, third, a somewhat lower average price for exports and higher average price for imports as foreign trade is governed more by the domestic supply situation and relatively less by foreign prices.

For all countries, the decrease in price variability is initially accompanied by a decrease in farm-income variability. The least-cost method of price stabilization moves prices in the opposite direction from that of production, and thus there is no conflict between price stability and income stability over a wide range. Once the coefficient of variation in price is reduced below the coefficient of variation of production, however, the coefficient of variation of farm income must increase, as it does for each country.

Average stocks increase as expected but never to high levels. Zimbabwe's average stocks—the highest of the three countries—are less than 250,000 tons, even for the most stable price regime in Table 8.3.

Characterizing the Optimal Policies

These average outcomes for all years do not indicate what types of policies are optimal. To study this in more detail, it is useful to

---

[13] The rapid inflation rate in Zambia over the last several years makes a similar calculation for that country highly suspect.

examine first a particular 10-year cycle to see how the optimal policies adjust price and trade to differences in production, world price, and stocks. This is followed by a more general characterization of these policies through regression analysis.

One of the 500 10-year cycles from the simulation for Zimbabwe, with price and quantity outcomes for both the free market (policy 1) and a more stabilizing policy (policy 2), is presented in Table 8.4. Policy 2 corresponds to the eighth line for Zimbabwe in Table 8.3, which has a coefficient of variation of price of 18.5 percent.

Table 8.4 highlights more dramatically than the summary statistics in Table 8.3 the extreme variability of price under free markets for these countries. The domestic price more than triples between years 6 and 7, is cut in half between years 7 and 8, then more than doubles between years 8 and 9. These price changes correspond to large changes in consumption, which is 50 percent higher in year 2 than in year 9. There can be no question that such extreme variability would have high transactions costs in a country where 50 percent of calorie consumption is from maize; in addition, the impact on investment—and even political stability—could be large.

The analysis of the stabilizing policy is best considered in two stages: first, the domestic price intervention, then the allocation of the resulting supplies between closing stocks and exports. Consider domestic price intervention first. Many analyses of grain-marketing reform have suggested that the parastatal become a "buyer and seller of last resort," letting prices move freely between a maximum and minimum, but intervening to stop any movement of price outside that range. Clearly policy 2 is not such a policy, as maize is bought domestically when the price is between 82 and 94 and sold at a price between 116 and 123. Consideration of more cycles would show a considerably wider range of intervention prices. The policy in effect is "leaning against the wind," every year moving prices somewhat closer to the target than free markets, taking account of the cost of doing so. The price-band policy implicitly assumes that movements of price between the limits are costless to society, while moving from 1 percent below to 1 percent above the maximum is infinitely costly. Since the costs of price instability outlined above are unlikely to accrue in that fashion, a price-band policy cannot be the least-cost method of price stabilization.

On allocating supplies between stocks and trade, it is clear that both the free-market policy and the stabilizing policy export to the profitable regional market prior to holding any interannual supply stabilization stocks. Indeed, both policies export exactly the maximum

Table 8.4—One 10-year cycle from the simulation of a stabilizing policy for maize in Zimbabwe

Year	World Price	Production	Opening Stock		Net Purchases		Consumption		Domestic Price		Net Commercial Imports		Food Aid		Marketing Profits	
			(1)	(2)	(1)	(2)	(1)	(2)	(1)	(2)	(1)	(2)	(1)	(2)	(1)	(2)
	(US$/metric ton)		(1,000 metric tons)						(US$/metric ton)		(1,000 metric tons)				(US$ million)	
1	109	2,149	100	100	143	359	2,006	1,790	62	90	-120	-120	0	0	8.7	-20.2
2	112	2,649	123	339	527	826	2,122	1,823	51	85	-201	-615	0	0	-13.6	-41.2
3	129	2,684	450	550	677	897	2,007	1,787	62	91	-819	-1,087	0	0	17.1	-7.5
4	102	1,826	307	360	-138	3	1,964	1,823	66	85	-120	-120	0	0	27.6	13.4
5	92	1,414	50	243	-70	-246	1,484	1,660	168	116	21	3	0	0	7.9	28.0
6	90	2,230	0	0	214	393	2,016	1,837	61	83	-120	-120	0	0	2.7	-21.3
7	105	1,271	94	273	-167	-362	1,438	1,633	187	123	31	89	43	0	24.7	26.5
8	98	1,913	0	0	120	143	1,793	1,771	90	94	-120	-120	0	0	8.3	5.2
9	93	855	0	23	-574	-775	1,429	1,630	191	123	277	466	297	286	54.1	4.2
10	95	2,273	0	0	246	432	2,027	1,841	60	82	-120	-120	0	0	1.0	-24.5
11	...	...	126	312	...	...	...	...	...	...	...	...	...	...	...	...

Note: Policy (1) is the free-market policy, while policy (2) is the policy with a coefficient of variation of price of 18.5 percent.

to this market—120,000 tons—in five out of ten years. Once that market is saturated, stocks are held up to a maximum (which varies with the world price), after which the remaining supply is sold in the world market. Years 2 and 3 show that this maximum stock is higher for the stabilizing policy than for the free-market policy. Nevertheless, exports to the world market are larger under the stabilizing policy than under the free market because consumption is stabilized at a much lower level in those years. Occasionally, the higher stocks yield a handsome payoff for the stabilizing policy, as for example in year 5 when large stocks carried over from year 4 are sold during a bad year at considerable profit. But in contrast, the stocks carried over from year 1 end up being quite costly, as bumper crops follow in the two subsequent years and all of that stock ends up being exported at low world prices. The higher stocks for the stabilizing policy also lead to somewhat lower receipts of food aid in years 7 and 9.

Marketing agents make money in nine out of ten years in this cycle under free markets, but lose money in five of these years and on average under the stabilizing policy. Clearly, most of these losses would accrue directly to the government marketing body or indirectly through explicit government subsidies, since private agents would not voluntarily hold interannual stocks under the stabilizing policy without some additional incentive.

The examination of one cycle gives some insight into the nature of these policies, but little guidance into exactly how to implement a low-cost, stabilizing policy. Regression analysis can help to characterize the policies if relationships between variables are approximately linear. As it turns out, the relationship between the state variables (production, world price, and stocks) and the domestic market control variable, net purchases, is approximately piecewise linear, as suggested by Figure 8.1. The figure shows that the net purchases function is flatter around mean production than for either high or low production. This is logical, implying that the initial production fluctuations are buffered less than large production fluctuations. Based on this figure, three separate regressions are estimated for each policy, with net purchases as the dependent variable and production, stocks, and world price as the three independent variables. The data are partitioned by production, with low and high production years separate from years that are approximately normal. The decision rule for the other control variable, net foreign trade, is simpler. All of the optimal policies, including free-market policies, use a similar rule: if "available supply"—that is, opening stocks plus net purchases—is negative, import that amount. Never import to build up stocks. If available supply is greater than zero but

Figure 8.1—Net purchases versus production, Zimbabwe

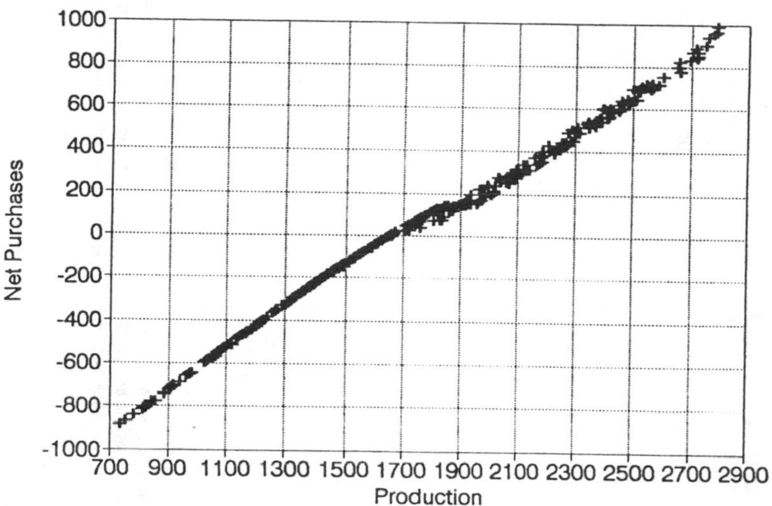

Note: World price is between 90 and 120; stocks are between 0 and 100; and coefficient of variation of price is 18.5.

less than the maximum regional export, export all of it and carry out no stock. If available supply is greater than the maximum regional export, hold stock up to a maximum, then export the remainder to the world market. This maximum stock depends on the world price.

The changes in consumption and price estimated by the regression equations, along with the maximum stock levels at two different world prices, are presented in Table 8.5.[14] For simplicity, only two policies are presented for each country, one that only slightly stabilizes prices and one that substantially stabilizes prices. Comparisons are best made between the less-stabilizing and more-stabilizing policies for the same country, since production and consumption levels and variability are quite different across countries. Several general patterns hold for all policies. First, as suggested by Figure 8.1, there is less attempt to

---

[14] Actual regression results are available from the author on request. As the figure implies, the fit is very good, with no t-statistic less than 9. More important, simulations of these policies yield results very similar to simulations of the interpolated optimal policies.

Table 8.5—Optimal policy adjustments to changes in stocks, production, and world price

Country/ Production Levels (PD)	Less-Stabilizing Policy: Change in Consumption with an Increase of 100,000 Metric Tons — Production	Less-Stabilizing Policy: Change in Consumption with an Increase of 100,000 Metric Tons — Stocks	Less-Stabilizing Policy: Change in Price with US$10 Increase in World Price	Less-Stabilizing Policy: Maximum Stock — World Price = US$80	Less-Stabilizing Policy: Maximum Stock — World Price = US$160	More-Stabilizing Policy: Change in Consumption with an Increase of 100,000 Metric Tons — Production	More-Stabilizing Policy: Change in Consumption with an Increase of 100,000 Metric Tons — Stocks	More-Stabilizing Policy: Change in Price with US$10 Increase in World Price	More-Stabilizing Policy: Maximum Stock — World Price = US$80	More-Stabilizing Policy: Maximum Stock — World Price = US$160
(1,000 metric tons)	(1,000 metric tons)		(US$/ metric ton)	(1,000 metric tons)		(1,000 metric tons)		(US$/ metric ton)	(1,000 metric tons)	
Malawi										
PD < 1,270	46	58	1.1			24	24	0.6		
1,270 <= PD <= 1,380	87	79	0.3			31	24	0.4		
PD > 1,380	27	38	1.3	300	20	6	9	0.8	140	60
Zambia										
PD < 1,200	24	44	3.2			12	21	1.5		
1,200 <= PD <= 1,320	73	54	1.1			34	21	1.2		
PD > 1,320	25	27	2.6	>680	100	11	14	1.7	540	180
Zimbabwe										
PD < 1,650	14	43	4.1			8	19	1.8		
1,650 <= PD <= 1,910	72	60	1.1			27	25	0.9		
PD > 1,910	27	32	3.4	1,230	60	9	10	2.4	870	150
Region										
PD < 4,000	35	57	3.2			19	28	1.5		
4,000 <= PD <= 4,400	82	77	0.4			37	31	0.9		
PD > 4,400	34	41	2.4	>1,400	50	12	14	2.1	1,400	450

Notes: The coefficients of variation of price for the two policies here are Malawi—16.2 percent, 6.1 percent; Zambia—28.4 percent, 14.1 percent; Zimbabwe—40.3 percent, 18.5 percent; region—36.1 percent, 17.0 percent. Domestic price adjustments to the world price are calculated one standard deviation of price below the mean, at the mean, and at one standard deviation above the mean price for low, medium, and high production, respectively. Stock figures preceded by a ">" indicate that stocks of that size are held but no exports are made to the world market even at the highest production level.

buffer production fluctuations for changes near mean production. Nevertheless, even the slightly stabilized policies do "lean against the wind," by removing 13,000-28,000 tons of a 100,000-ton increase in production close to the mean. Considerably more is removed when production is far from the mean. The more-stabilizing policies show the same pattern but buffer all production fluctuations more completely. Once again, these policies clearly are not price-band policies.

The pattern of responses to the world price is quite different, with the response when production is close to the mean smaller than when production deviates from normal. Since few exports and imports take place when production is close to normal, the importance of adjustments to the world price is minimized. None of the responses to the world price are particularly large, especially for the more-stabilizing policies.

Maximum stock levels respond to a stronger desire for price stabilization in an interesting way. At low world prices maximum stocks decrease, while at high world prices they increase. This is a result of the different pattern of consumption and net purchases as price becomes more stable. The marketing body must buy and sell small quantities more frequently under a stabilizing policy, and must import if necessary even when world prices are high. Thus small amounts of stock are more likely to displace imports, and consequently more are held when world prices are high. When world prices are low, however, stocks are considerably higher and, on the margin, unlikely to be used to displace future imports. Under a less-stabilizing policy, there is a good chance that the stock will be consumed in the subsequent year (column 2 in Table 8.5). As shown in column 7 of the table, however, once stocks are put into storage under a more-stabilizing policy, they frequently cannot be sold domestically without endangering the price stabilization objective and are thus likely to incur storage charges for several years. So exports to the world market, even at very low prices, are more attractive than large stocks. Average stock levels increase with price stabilization, as shown in Table 8.2, not because of these changes in maximum stocks but because purchases in high production years more frequently are larger than the maximum regional exports.

In sum, there are five key adjustments to policy that arise from this analysis of optimal policies. The first three concern domestic prices, which move in response to world price, domestic production, and stocks. The exact response varies with production and the importance of the price-stability goal, but all optimal policies include these three adjustments. The last two adjustments concern stocks and trade: no interannual supply stabilization stocks are held until the profitable regional market is saturated; once it is, stocks are held until a maximum is reached, which varies with both the world price and the importance of the price-stability goal.

FROM MODELING TO REALITY: IMPLICATIONS FOR POLICY DESIGN

All models, and optimizing models in particular, must simplify reality in order to produce usable results. A model suggests ways that policies might be changed and presents ballpark figures for how a particular policy change might affect variables that concern government. Models therefore provide a starting point for a discussion of policy reform; they do not and cannot prescribe the perfect policy to implement.

This section examines implications of the modeling effort for domestic price policy,[15] foreign trade, and public stockholding.

Implications for Price Policy

Many less developed countries set the price of the staple food prior to any knowledge concerning the size of the crop, and then maintain that price until the next market year. On the surface, therefore, these policies stabilize prices completely within years and substantially between years (although in reality official price stability has frequently been consistent with high price fluctuations on parallel markets). The optimal policies, on the other hand, move the domestic price in response to world price and domestic production. Two questions arise here concerning implementation: how much should the domestic price move in response to these variables, and how can the government determine movements of this size?

Since the benefits of price stability have not been measured here, there is no clear answer to the first question. The very low welfare cost of decreasing the coefficient of variation of price to 16 percent in Malawi and 28 percent in Zambia indicates that at least that degree of stabilization is warranted in those countries, while the exceptionally high variability found under free markets in Zimbabwe and for the region again suggests that the benefits of some stabilization should be large. Stabilizing prices beyond the point where farm income is most

---

[15] See Pinckney 1989 for a more detailed discussion of this issue.

highly stabilized, however, is less likely to be worth the cost, as some of the benefits of stabilization accrue because of farm-income stability rather than price stability. In the absence of precise measures of the benefits, the trade-offs between price stability, income stability, welfare, and marketing cost as implied by Table 8.3 may be useful in the political decisionmaking process. For example, decreasing price variability from 26.9 percent to 11.2 percent in Zimbabwe costs about US$11.4 million annually in marketing costs (most probably paid by the government), an additional US$l million annually of consumer surplus, and an increase in farm-income instability of about 3.4 percent.

There are at least two ways that a more responsive policy could be built into existing parastatal marketing systems. The first method would change the price at which the parastatal buys and sells at more frequent intervals. The parastatal would then allow its prices to move in the same direction as market prices, but to a lesser extent. As information accrued concerning the size of the domestic harvest and the international situation, the parastatal would change its prices accordingly. To minimize the potential for corruption, it might be best to make such price changes depend as much as possible on known, readily accessible, published data. This could easily be done for the international market price, but would be more difficult domestically until regional markets and market-information systems are well developed. The more frequent the price adjustment, the smaller will be each movement, thereby minimizing the incentive for profitable inside information from leaks. Although this procedure would be quite different from the existing, once-a-year price review, the process is similar to the daily setting of foreign-exchange rates by central banks, which is practiced successfully by many countries in the region.

An alternative that would change the nature of government intervention but might be less susceptible to corruption would be a system of tenders. The government marketing agency would make regular tender offers of specific quantities of maize for sale or purchase. The price at which the tenders are bought or sold would provide the government with valuable market information to be used in conjunction with information from other sources in determining the quantities offered the following period. The government would then try to affect prices by interventions in the quantity available locally, rather than by trying to influence the price directly.

The best method for dampening price fluctuations will vary from country to country, and in the same country over time. But the least-cost policy always will include some degree of "leaning against the wind," thus dampening price movements in most years while allowing official prices to move in the direction that market pressures indicate.

Implications for Foreign Trade

The implications of the model for foreign trade are straightforward: export as much as possible to the profitable regional market, then export to the world market only after the (world pricedependent) maximum stock level is reached.

These implications clearly depend on the assumption concerning the price at which food-aid donors will buy maize in the region. That assumption—$120 above export parity to the world market—is consistent with past behavior but is obviously subject to change. The assumption concerning the amount that could be exported to this market by each country is purposely on the low side in the model; clearly, in the real world the amount would vary from year to year and, it is hoped, will decline over time as civil conflicts decrease in the region. Nevertheless, these markets are clearly more profitable than holding interannual supply stabilization stocks under a wide range of parameter values; the recommendation to sell to this market before holding such stocks is robust.

The model makes no assumption about whether imports and exports are handled by private traders or by a government parastatal; the assumption is only that government can control the amount of foreign trade. Two aspects of the real world, not featured in the model, would suggest that at least some trade in small lots should be allowed by private agents across borders. First, a ban on private trade in small lots is costly to enforce, and those enforcement costs constitute welfare losses to the country. Second, as Nuppenau shows (see Chapter 12), the regional pattern of production leads to numerous opportunities for profitable trade across borders; in some cases, it makes sense for two countries to both import from and export to each other in the same market year. It is unlikely that a government parastatal could ever take advantage of these numerous, small opportunities for profitable trade.

But in order to allow such trade while still maintaining control of overall trade volumes, there must be some combination of clear guidelines for what trade can take place without government interference and what trade requires approval. Small lots—which in total would not heavily influence the national food situation—could be allowed in and out without interference. Larger lots could be subject to trade taxes or subsidies as the situation warrants, with the amount of the tax or subsidy changing as domestic and international market conditions change.[16] Such trade taxes or subsidies are more desirable than approving a set number of foreign contracts and disapproving the remainder, for two reasons. First, the effect of the trade taxes on domestic prices is clear, while the impact of the quantitative restrictions could be quite different from what is expected. Second, approval of a few foreign contracts and disapproval of others encourages corruption and inefficiency, as those who are awarded the contracts are likely to make large profits.

The model examines the benefits of completely free trade in maize between Zambia, Zimbabwe, and Malawi by aggregating the three into one country. Model results give little support to the idea that a regional free-trade zone would help stabilize maize supplies and prices, although these results are subject to significant caveats. Because production fluctuations are fairly highly correlated in the region, regional production fluctuates almost as much as the average of the three countries. Consequently, reducing price fluctuations costs as much at the regional level as for the sum of the different countries. Malawi, in particular, would appear to suffer from a regional grouping, as it would move into a considerably higher-variability price regime.

The model thus shows that regional free trade is unlikely to assist these countries in dealing with national, interannual supply stabilization issues. However, there could be substantial benefits of such a policy resulting from allowing trade across borders between adjacent surplus and deficit regions, as discussed above.

Implications for Stockholding

The model suggests that no interannual supply stabilization stocks should be held in most years, and that regional exports should take priority over stockholding. Is this reasonable in the real world?

Note that this policy recommendation does not imply that no stocks would be held by the government during those years. Depending on the seasonal price policy, seasonal stocks may be held within the market year, and additional stocks are necessary to ensure that the market functions normally between the time that the need for imports is recognized and the arrival of those imports in the country. Although this latter type of stock—termed "import buffer stocks" in Pinckney

---

[16] In the context of the methods of domestic price intervention discussed above, these trade taxes would have to change with each resetting of the domestic price or with each tender offer.

1989—has a large seasonal component, a prudent policy will usually lead to some of these stocks being carried over from one market year to the next.[17] Nevertheless, the model does imply that stocks in many years would be far below historic levels.

There are three possible objections to this recommendation: the possibility of speculative attack, the lack of foreign exchange to purchase imports, and the unavailability of maize from the world market. These will be considered in turn.

In a year of scarcity, private agents may buy all the maize in government warehouses before imports arrive, thus driving up prices and reaping large profits. Such behavior has been termed "speculative attack" in the literature. Government does need to be concerned about the possibility of speculative attack. Note two points, however. First, this issue concerns import buffer stocks, not interannual supply stabilization stocks. If there is a problem, it results from insufficient supplies to provide for the time between the recognition of the need for imports and their arrival in the country, not because of insufficient supplies to buffer production fluctuations. The model strongly suggests that the year after maize is imported, production fluctuations should be buffered by trade alone, and not by stocks. Second, the best way to combat incentives for speculative attack is through consistent government policy, with the government living up to promises. If the populace has confidence that government is arranging to control the supply situation and the marketing body continues to sell to everyone at the announced price, price expectations should be stable and there will be no incentive to store. Consistency of government policy is the key to building confidence of the people in the ability of the government to control supply, and this confidence results in low expectations of future price increases.

One reason for such speculative attack would be if the public doubted the ability of the government to purchase imports because of foreign-exchange availability. All of the policies considered here earn foreign exchange, as shown in Table 8.3, because of the profitability of regional exports. A policy that holds stock rather than exporting to the regional market actually damages the country's foreign-exchange position. One plausible procedure for guaranteeing the availability of foreign exchange for imports and ensuring public confidence would be

---

[17] See Pinckney 1990 and Duncan et al. 1990 for analyses of the size of import buffer stocks in Malawi and Zimbabwe, respectively. A similar analysis for Zambia would be timely.

to keep some of the proceeds from maize exports in a blocked, interest-bearing account denominated in foreign exchange. Decisions could be made ahead of time about the circumstances in which imports would be authorized. This would cut down on delays inherent in approval of foreign-exchange allocations and high-level decisionmaking about imports.

The final concern about low stocks would be the unavailability of maize on the world market. For yellow maize, there is no issue; even in the midst of the world food crisis in the early 1970s, maize could always be bought internationally. White maize is more problematical, as there have been periods in the past when the crop was unavailable. If consumers prefer white to yellow maize but white maize is unavailable, demand can be equilibrated with supply by allowing yellow maize to sell at a substantial discount. Model results not reported here suggest that recommendations for stockholding change very little if imported maize sells at a discount of US$25 per ton.

Nevertheless, this study does not recommend low stocks in all years; instead, interannual supply stabilization stocks should vary with world price, opening stocks, and production, as shown above. If Zimbabwe had been pursuing a policy of this type in the mid-1980s, when production was high and world prices were at historic lows, it would have held large stocks. For example, closing interannual supply stabilization stocks on March 31 of 1986 and 1987 would have been between 0.7 and 0.9 million tons, assuming prices were stabilized to a coefficient of variation between 6 percent and 19 percent. Given that a prudent import buffer stock policy could lead to additional interannual stocks of about 0.2 million tons, total closing stocks in those years would have been between 0.9 and 1.1 million tons, compared to actual stocks of 1.4 million tons in 1986 and 1.8 million tons in 1987. These recommendations for stock levels in the mid-1980s are considerably higher than those reported in Buccola and Sukume 1988, although both studies suggest that government stocks were excessive in this period.[18]

One possible reason for the reluctance of a government, including Zimbabwe's, to export to the world market at such times is that substantial losses would accrue in the parastatal's books by selling stock

---

[18] The Buccola and Sukume (1988) analytical model differs substantially from this model; the primary determinants of the difference in recommended stock levels appear to be (1) that Buccola and Sukume undervalue stocks by limiting the time considered to two periods, and (2) that the model in this study overestimates optimal stockholdings by ignoring supply response. See footnote 8.

below its purchase price. Accountants value stocks at the purchase price. But the price paid is not the economic value of the stock. If the stock is not to be used in the present market year, its value is a weighted average of the different possible uses next year (with each weight equal to the probability of each possible use), divided by a discount rate, minus the cost of storage. If this value is less than the net proceeds from exports, the stock should be exported. Thus, at times it is profitable for the country to export at an accounting loss.

CONCLUSIONS

Countries around the world are reexamining past policies that were not supportive of market agents and are making numerous adjustments. Markets for staple foods are some of the most difficult to adjust because of the legitimate government goal of stabilizing real incomes, even as the government tries to encourage efficient investment, production, and trade. Although past government policies generally have been costly, while, in many cases, failing to stabilize parallel market prices, a system of laissez-faire is unlikely to be an improvement, since a high level of production instability and large differences between import and export parity would lead to wildly fluctuating staple food prices. This study has explored low-cost methods of price stabilization, and suggested general policy rules for achieving this goal. These rules generally move the country in the direction of free markets from present policies but maintain an important role for a government marketing organization in pushing prices toward stability in all years.

Some features of the model—particularly the two-tiered export market, with a profitable market for small lots in all years—are peculiar to the region, and recommendations concerning that market are clearly not generalizable. But the similarity of results for the countries considered here strongly suggests that all low-cost price-stabilization policies for countries that move between an export and import position will include the major features of domestic price intervention, foreign trade, and stockholding discussed here.

REFERENCES

Bigman, D. 1985. *Food policies and food security under instability: Modeling and analysis*. Lexington, Mass., U.S.A.: Lexington Books.

Buccola, S. T., and C. Sukume. 1988. Optimal grain pricing and storage policy in controlled agricultural economies: Application to Zimbabwe. *World Development* 16 (3): 361-371.

______. 1991. Regulated-price and stock policies: Interaction effects and welfare preference. *Journal of Development Economics* 35 (2): 281-305.

Dawe, D. 1990. Stable food prices in a macro growth model. Harvard Institute for International Development, Cambridge, Mass., U.S.A. Mimeo.

Duncan, A., J. Gray, W. Masters, and R. Pearce. 1990. Background paper for the World Bank Agricultural Sector Review of Zimbabwe. World Bank, Washington, D.C. Mimeo.

Gustafson, R. L. 1958. Carryover levels for grains: A method for determining amounts that are optimal under specified conditions. Technical Bulletin 1178. Washington, D.C.: U.S. Department of Agriculture.

Jayne, T. S., and M. Chisvo. 1991a. Unravelling Zimbabwe's food insecurity paradox: Implications for grain market reform in southern Africa. *Food Policy* 16 (4): 319-329.

______. 1991b. Zimbabwe's grain marketing policy challenges in the 1990's: Short run vs. long run options. Paper presented at the Seventh Annual Conference on Food Security in Southern Africa, October, Victoria Falls.

Jayne, T. S., M. Chisvo, S. Chigume, and C. Chopak. 1990. Grain market reliability, access and growth in low-potential areas of Zimbabwe: Implications for national and regional supply coordination in the SADCC region. In *Food security policies in the SADCC Region*, ed. M. Rukuni, G. Mudimu, and T. S. Jayne, 113-127. Harare: University of Zimbabwe.

Koester, U. 1986. *Regional cooperation to improve food security in southern and eastern African countries.* Research Report 53. Washington, D.C.: International Food Policy Research Institute.

Massell, B. F. 1969. Price stabilization and welfare. *Quarterly Journal of Economics* 83 (2): 284-298.

Newbery, D. M. G., and J. E. Stiglitz. 1981. *The theory of commodity price stabilization: A study in the economics of risk.* Oxford, U.K.: Clarendon Press.

Oi, W. Y. 1961. The desirability of price instability under perfect competition. *Econometrica* 29 (January): 58-64.

Pinckney, T. C. 1988. *Storage, trade, and price policy under production instability: Maize in Kenya.* Research Report 71. Washington, D.C.: International Food Policy Research Institute.

_____. 1989. *The demand for public storage of wheat in Pakistan.* Research Report 77. Washington, D.C.: International Food Policy Research Institute.

_____. 1990. *The design of storage, trade, and price policies for maize in Malawi.* Williamstown, Mass., U.S.A.: Williams College for Harvard Institute for International Development.

Samuelson, P. A. 1972. The consumer does benefit from feasible price stability. *Quarterly Journal of Economics* 86 (August): 476-493.

Stiglitz, J. E. 1987. Some theoretical aspects of agricultural policies. *The World Bank Research Observer* 1 (1): 43-60.

Timmer, C. P. 1989. Food price policy: The rationale for government intervention. *Food Policy* 14 (1): 17-27.

_____. 1990. *Food price stabilization: The Indonesian experience with rice.* Cambridge, Mass. U.S.A.: Harvard Institute for International Development for U.S. Agency for International Development.

van der Geest, W. 1991. Food security: The cereal market policy model, a non-technical introduction. Food Studies Group Working Paper No. 3. Oxford, U.K.: International Development Centre.

Waugh, F. V. 1944. Does the consumer benefit from price instability? *Quarterly Journal of Economics* 58 (August): 602-614.

Wright, B. D., and J. C. Williams. 1988. Measurement of consumer gains from market stabilization. *American Journal of Agricultural Economics* 70 (August): 616-627.

9
Maize Movement and Pricing Decontrol in Zimbabwe

Thomas S. Jayne and Ernst-August Nuppenau

Uncertainty over the effects of grain price and movement decontrol are likely to circumscribe the pace and extent of the grain-market reform program in Zimbabwe. While the government of Zimbabwe is increasingly aware of the costs of its grain policies in terms of budget deficits and food insecurity, it may nevertheless be reluctant to implement major changes without detailed analysis of their effects on selected socioeconomic groups and government objectives. Recent experiences with grain-market liberalization in other African countries demonstrate that the risks of miscalculation can be extremely high. In Zimbabwe, major uncertainties and conflicting perceptions prevail within the relevant ministries and at the highest levels of government concerning the effects of market decontrol on (1) the direction and magnitude of changes in producer and consumer grain prices; (2) the trading account of the Grain Marketing Board (GMB); (3) the GMB's ability to procure adequate supplies to perform supply stabilization functions; and (4) the ability of the private sector to develop competitive and efficient trading channels within a restructured marketing system.

How would decontrol of maize movement and pricing affect regional prices and producer and consumer welfare, the GMB's trading account, and national maize self-sufficiency? These issues are analyzed within a dual-market, spatial-equilibrium model that explicitly considers interactions between the official and informal trading and milling sectors. Three scenarios are examined:

1. The current system of grain movement restrictions and panterritorial prices.
2. Relaxation of grain-movement restrictions between all smallholder areas (not commercial areas) and between smallholder and urban areas, while still maintaining panterritorial prices for state-traded maize.
3. Full relaxation of grain-pricing and movement controls in smallholder and commercial areas.

Results indicate that deregulation of maize movement—even given continuation of current price controls—would substantially alter consumer and producer prices in many smallholder areas as well as major urban centers. The model indicates that deregulation of maize produced in smallholder areas causes changes in trade flows that result in lower maize-meal prices in urban areas and higher producer prices in smallholder areas near urban centers. Gross farm incomes in surplus smallholder areas would rise, while grain prices in food-deficit areas would fall, thus promoting food security. While GMB intake from smallholder areas would decline moderately, demand for GMB maize would also decline, resulting in a slight increase in exportable surplus. The GMB's domestic trading losses would fall moderately because of a shift in the relative proportion of intake from high-cost smallholder areas to lower-cost commercial areas.

If movement decontrol is extended to commercial farmers as well, the results are substantially different: the commercial sector, because of lower transport costs to urban centers, generally replaces smallholder areas in fulfilling urban demand. Producer prices rise in commercial farming areas and decline in most high-productivity smallholder areas, except those close to urban centers and along export routes. These findings are consistent with existing perceptions that due to their proximity to major consumption centers, commercial farmers stand the most to gain from movement decontrol (Grain Marketing Board 1991). However, the results suggest that total movement deregulation and regionally differentiated pricing at GMB depots would improve the GMB's domestic trading account. Moreover, the GMB would generate substantially larger national maize surpluses, due to a supply response in commercial and favorably located smallholder areas. Reform may thus offer an important benefit in the current environment of dwindling national maize supplies.

In the long run, Zimbabwe's food-price dilemma may be relieved by new farm technology, resettlement, the successful generation of employment and income growth, or a combination of these developments. However, these gains do not appear to be on the immediate horizon, especially in Zimbabwe's semi-arid areas, where the majority of smallholders live. In the short and medium runs, efforts to reduce marketing costs through development of informal distribution and milling may simultaneously raise producer prices and reduce maize-meal prices for low-income consumers.

ZIMBABWE'S MAIZE-MARKETING SYSTEM: POLICY OBJECTIVES AND CURRENT PERFORMANCE

Since independence in 1980, some of the main stated agricultural policy objectives of the government of Zimbabwe have been income growth among rural smallholders; food security, with particular attention to the urban and rural poor; and the minimization of budgetary losses arising from government marketing and pricing operations (Zimbabwe 1982, 1988).

Maize marketing and pricing policies have been primary instruments for achieving these broad objectives.[1] The expansion of GMB infrastructure into the communal lands was a pillar of postindependence policy to promote income growth among smallholders. GMB expansion into the more distant and arid areas is often referred to as a "social function" because it was not justified on strictly commercial terms. The government also pursued its policy objectives by maintaining stable and remunerative GMB producer prices and by expanding smallholders' access to government credit, recouped from crop sales to the GMB. These policies contributed to the dramatic rise in GMB grain intake from the smallholder sector.

Zimbabwe's grain-marketing system facilitates a number of important and often unrecognized transfers of income between groups that are inconsistent with stated government policy objectives. These income transfers occur through explicit subsidies but, often more important, also through regulations and policies inherent in the organization of the marketing system.[2]

Major Problems of the Grain-Marketing System

1. Regulations block grain from moving directly from surplus to deficit rural areas. As a result, most surplus grain production is channeled into the GMB/urban commercial milling system. This creates a circuitous rural-urban-rural flow of grain through a high-cost milling system in order to meet rural demand. The system perpetuates a wasteful use of transport and artificially high consumer prices, exacerbating food insecurity.

2. The unidirectional GMB system, while providing clear benefits to surplus producers, cannot cost-effectively distribute grain to

[1] The importance of maize in the agricultural economy of Zimbabwe is apparent from the following: maize accounts for 88 percent of coarse-grain production in the country, 80 percent of coarse-grain production among smallholders, and 61 percent of coarse-grain production among smallholders in Natural Regions 4 and 5 since 1981 (AGRITEX, various years). Furthermore, the proportion of smallholder area and production devoted to maize appears to be gradually increasing (AGRITEX, various years). Maize meal accounts for 45 percent of the total caloric intake in the Zimbabwean diet (USDA 1988).

[2] For more details on the grain-marketing system see Appendix 1.

geographically dispersed and remote areas. The GMB system is well-suited for high-volume intake and distribution, but cannot operate cost-effectively where geographically dispersed rural demand precludes economies of scale in distribution. The use of GMB depots as retail outlets makes economic sense only for consumers within close proximity to these depots. Underdeveloped informal trading networks create a situation in which GMB stocks in town centers are largely inaccessible to consumers in remote rural areas.

3. Because grain cannot be transported informally from surplus to noncontiguous, deficit rural areas, the system places increased emphasis on the industrial milling system to meet rural demand during drought years. This transfers income from grain purchasers and rural small-scale millers (along with any multiplier and employment effects) to urban industrial millers. The phenomenon of increased demand for urban-milled meal during drought years is largely due to the failure of the marketing system to allow more direct redistribution of grain from surplus to deficit smallholder areas.

4. The share of the maize-meal price accruing to producers has declined over the past decade. The producer received 44 percent of the full cost of roller meal (including subsidies) in 1991/92, compared with an average of more than 50 percent during the early and mid-1980s. Therefore, social functions that inflate GMB margins at a time when subsidies are to be cut must come at the expense of lower real producer prices or higher real consumer prices, or both. Over the past decade, the government of Zimbabwe has chosen to extract the cost of these social functions out of the producer price (Jayne and Chisvo 1991), contributing to an erosion of the national maize-production base. Commercial maize area is declining at an annual rate of 18,000 hectares a year (90,000 metric tons a year, given average yields). Meanwhile, smallholder maize sales to the GMB peaked in 1985/86.

5. The system encourages a pattern of regional self-sufficiency in grain production inconsistent with comparative advantage and income growth in the semi-arid areas. Inflated consumer grain prices encourage grain production for home consumption in low-rainfall areas and discourage diversification into higher-valued oilseed crops that generate foreign exchange (Jayne et al. 1991). Policies that have raised acquisition prices for grain in rural areas work against government objectives in two ways, by inflating the amount of income spent on food and by lowering the value of farm output sold. Cheaper and more reliable access to grain in the drier areas is necessary to reduce households' overriding concern with grain self-sufficiency and to promote dynamic changes in crop mix more consistent with compara-

tive advantage and income growth in the low-rainfall areas.

6. Low-income urban consumers pay artificially high prices for maize meal due to controls on grain movement and resale that restrict informal traders' and millers' access to grain. Considering that urban unemployment levels currently stand at 30 percent, artificially high prices of the most important staple food in the Zimbabwean diet erode household food security among the poor.

Perhaps the greatest difficulty with the present organization of the marketing system is that it is increasingly unsustainable. Without changes in pricing and market regulations, the GMB has two options: either offer attractive producer prices and subsidized selling prices—that is, a subsidized GMB margin—to capture most of the marketed maize surplus and incur major budgetary losses in the process, or widen the GMB margin to cover costs and lose market share on the GMB's profitable trading routes. Attractive opportunities already exist for private maize trading that bypasses the GMB in areas where transport costs between producer regions and urban centers are low. A further widening of the GMB margin by 30-40 percent to eliminate subsidies will create additional incentives for private trade that contravene existing market regulations. The inability of most developing countries to suppress illegal informal trade when state regulations are no longer compatible with producer or consumer interests suggests that the existing system is becoming increasingly unsustainable. Apart from the desirability of market reforms, changes will become imperative in an environment of GMB subsidy reduction.

MODELING TRADE FLOWS IN A DUAL MARKETING SYSTEM

The foregoing suggests that quantitative estimates of the impact of maize-marketing reforms will require a model that captures interactions between the official and informal maize-marketing systems. The results of this analysis are based on a dual-market, regional, spatial-equilibrium model. The model is used first to examine formal and informal trade flows, net trade, price levels, and the GMB trading account under the current set of policy restrictions on grain movement and prices. The results from this base case (Scenario 1) are compared with those from two alternative scenarios: allowing grain trade from smallholder to urban and commercial areas (Scenario 2), and allowing for free grain trade between all regions at equilibrium-determined prices (Scenario 3). The GMB may still maintain a role in Scenario 3 by buying and selling at market-determined prices. External trade is assumed to remain in the hands of the state in each scenario.

The model is essentially structured as follows. Maize supply functions are estimated econometrically for each producing region (13 smallholder areas and 5 commercial areas). GMB producer price, chosen exogenously in Scenarios 1 and 2, determines GMB intake and influences the supply of grain in informal markets. The informal price, which is derived from local supply and demand conditions in each smallholder area, is nevertheless influenced by government pricing decisions in the official market. The government-determined price of industrial maize meal serves as a ceiling price in the informal market in Scenarios 1 and 2, but becomes unregulated (endogenous) in Scenario 3. When movement restrictions are in force, each region is in autarky, except for the movement of industrially milled meal to meet demand in deficit regions. When movement restrictions are relaxed, the model is similar to standard trade models where excess supply and demand curves are determined from the supply and demand curves in the respective regions. Excess supply and demand determine a unique informal price in each region, which is modified by relevant transport and processing costs. Sensitivity analysis on these margins is possible to examine the robustness of trade flows and prices to various assumptions about the competitiveness and efficiency of the informal market. Finally, by aggregating across regions, national supply to the GMB and sales of industrially milled meal can be derived. The residual, after adjusting for milling extraction rates and demand for GMB grain from stockfeeders and brewers (which is a relatively small part of the market and treated as a constant), is national surplus, that is, endstocks plus net exports.

The analytical framework of the model and elasticities used to run the model are presented in more detail in Appendix 2.

RESULTS

Movement Decontrol of Maize Produced in Smallholder Areas Only (Scenario 2)

Model results suggest that decontrol of maize movement in smallholder areas and between smallholder and urban areas, while maintaining panterritorial pricing on GMB-traded maize, would result in the following:

- Low-income urban consumers pay lower prices for maize meal due to the ability of informal millers to procure maize directly from smallholder areas. As a result, urban maize consumption increases. Consumption shifts moderately from industrially milled to informally milled meal. The results indicate that 56,000 tons of maize flow into Harare and 19,000 tons into Bulawayo for informal milling, causing a decline of 48,000 tons and 16,000 tons, respectively, of industrial-meal consumption in these areas. This represents a 29 percent decline in total demand for industrial maize meal. The bulk of human consumption is still in the form of industrial meal because it is preferred by most middle- and high-income consumers.

- Lower milling costs in urban areas are passed on to smallholder areas near urban areas in the form of higher producer prices. As a result, total smallholder grain sales (GMB plus informal) increase by an estimated 16 percent. Because a greater portion of urban demand requirements is met through lower-cost informal channels, demand for industrial meal and, indirectly, for GMB maize declines by an estimated 15 percent (Table 9.1). The net GMB surplus (intake minus sales) is 4 percent higher than that of Scenario 1.[3]
- Informal producer prices in most of the drier smallholder areas rise moderately, as these areas also experience increased demand from urban centers. Since the price of industrial maize meal serves as a ceiling on acquisition prices in informal markets, there is little or no change in the marginal price of grain for consumers (Table 9.2).
- The volume of trade between smallholder areas is relatively small. Intrarural maize trade takes place from Gokwe/Hurungwe to Omay/Hwange and from Buhera/Gutu to Mberengwa/Chivi and Bikita/Zaka. However, the largest volume of smallholder trade is to Harare and other urban centers.
- Movement decontrol in smallholder areas shifts the relative proportion of GMB intake from commercial areas. Since the GMB incurs lower per-unit costs in these areas and typically generates a profit to cross-subsidize its operations in the higher-cost communal areas, this policy change is estimated to reduce the GMB domestic transport and handling costs by 13 percent.

---

[3] The effects of partial movement decontrol on the GMB's ability to meet national demand under a broader range of weather conditions needs further analysis.

Table 9.1—Estimates of maize trade flows and distributional effects resulting from selected policy reforms: normal rainfall case

Scenario[a]	GMB Intake			GMB Sales to Commercial Millers for Maize Meal			Net GMB Surplus[b]	Smallholder Maize Sales		Distributional Effects Relative to Scenario 1			
	Commercial Farmers	Smallholders	Total	Urban	Rural	Total		Informally Traded and Milled	Total (GMB plus Informal)	Urban Consumers	Commercial Farmers	Surplus Smallholder Areas	Deficit Smallholder Areas
	(1,000 metric tons)												
1	493	576	1,069	456	121	577	252	108	684	...	...	...	...
2	481	510	991	375	114	489	262	283	793	+	0	+	0
3	668	549	1,217	275	113	388	589	174	723	+	+	+/–[c]	0

Note: GMB is Grain Marketing Board.

[a] Scenario 1 is the existing case of grain-movement restrictions and panterritorial prices on state-traded maize. Scenario 2 is relaxation of maize-movement restrictions on smallholder maize, while maintaining panterritorial producer and maize-meal prices. Scenario 3 is full relaxation of maize movement on commercial and smallholder maize, and introduction of spatially differentiated prices according to supply and demand conditions.

[b] The surplus is calculated after subtracting demand for GMB maize from stockfeeders, brewers, and drought relief (assumed constant at 240,000 metric tons).

[c] Effects depend on location; smallholders closer to urban areas will receive higher farm incomes, while remote areas will feel little effect.

Table 9.2—Preliminary estimates of regional price changes for maize grain and meal resulting from selected policy scenarios: normal rainfall case

	Changes in Price Levels from Scenario 1			
	Scenario 2		Scenario 3	
Region	Producer[a]	Consumer[b]	Producer[a]	Consumer[b]
	(percent of change)			
Urban areas				
Harare	...	-21	...	-21
Bulawayo	...	-8	...	-16
High-potential smallholder areas				
Gokwe/Hurungwe	[c]	[c]	-5	-3
Chiweshe/Murehwa	+10	+6	+8	+6
Guruve/Dande	+12	+8	-10	-7
Grain-deficit smallholder areas				
Matabeleland South	+16	[c]	+15	+8
Mberengwa/Chivi	+3	-4	-6	-4
Bikita/Zaka	+12	-8	+15	+11
Omay/Hwange	[c]	[c]	-2	-1
Lupane/Tsholotsho	+15	[c]	-8	-5
Beitbridge	-3	-6	-4	-7
Marginally surplus or deficit smallholder areas				
Buhera/Gutu	+1	+1	-2	-1
Chimanimani	+4	+2	+3	+2
Nyanga	+14	+4	+14	+9
Mutoko/Mudzi	[c]	[c]	-16	-9
Commercial urban areas				
Mashonaland	[c]	-19	+13	-21
Midlands	[c]	-16	+24	-18
Masvingo	[c]	-14	+29	-17
Matabeleland	[c]	-7	+20	-9

[a] Producer prices refer to maize grain. In Scenarios 1 and 2, the fixed Grain Marketing Board producer price serves as a floor price.
[b] Consumer prices refer to maize meal through informal channels. In Scenarios 1 and 2, the fixed price of industrial maize meal serves as a ceiling price.
[c] No change.

Full Decontrol of Maize Movement and Pricing (Scenario 3)

Preliminary results suggest that full market decontrol, under average weather conditions, will have the following effects relative to the existing set of marketing policies:

- The relaxation of controls on maize movement into urban areas substantially increases consumption of informally milled meal in urban areas (Table 9.2). This meal is estimated to be about 20 percent cheaper than industrial roller meal (1991/92 prices). However, in contrast to Scenario 2, these supplies come primarily from nearby commercial farms. This is because the bulk of commercial maize production occurs in close proximity to Harare and other urban areas in the north. Commercial farmers also benefit from cheaper rail transport to urban areas. Only several smallholder areas continue to export substantial volumes to urban areas. Supplies to Bulawayo also come mainly by rail from Mashonaland commercial farming areas and to a lesser extent from nearby smallholder areas.
- The higher producer prices in commercial farming areas induce a supply response that increases total maize sales from this sector by 35 percent over Scenario 1. By contrast, total smallholder sales decline by about 5 percent. The GMB's net surplus (intake minus sales) rises to nearly 600,000 tons (Table 9.1). This result is noteworthy in light of stated concerns of officials at the GMB and the Ministry of Agriculture regarding the ability of the GMB to procure adequate supplies to maintain strategic stocks in a liberalized market. The GMB's ability to do so appears crucially dependent on its ability to set regionally differentiated prices that are in line with informal supply and demand conditions so that it captures the bulk of the surplus from the commercial sector. This is assumed to be the case in Scenario 3.
- The bulk of urban maize-meal consumption is still processed by the industrial millers because of preferences for the more refined meal among the majority of urbanites. Thus the industrial and informal milling sectors may be viewed as complementary. Each sector appears to fill specific niches in the maize-meal market.
- Due to the potential for the lower-cost informal milling system to operate in urban areas, lower marketing costs are passed on to urban consumers with a preference for informal meal and to smallholders in surplus areas close to these urban centers who receive higher producer prices. However, producer prices fall in several high-productivity smallholder areas by 5-10 percent (relative to Scenario 1) due to increased competition from commercial farmers (Table 9.2).
- As in Scenario 2, relatively little trade takes place from surplus to deficit smallholder areas. This is because of relatively high transport costs and poor road infrastructure linking smallholder

areas to one another. Intrarural commerce must often follow a "V-shaped" pattern requiring transport into towns in order to go back out to nearby rural areas. The results suggest that improved road infrastructure may be necessary to exploit potential gains from trade among smallholder areas.

- The effects of removing panterritorial prices appear to have a relatively small effect in comparison with movement decontrol. This is because the existing official marketing margin (that is, GMB margin plus industrial milling margin) is so large that informal market-determined prices compare favorably with the floor and ceiling price bands of the existing official marketing system. Most producer and consumer prices in Scenario 3 fall within this price band.
- Consumer prices for maize meal do rise, however, by up to 11 percent in selected grain-deficit smallholder areas. This would have an adverse effect on food security in these areas.
- Because of generally lower consumer prices in this scenario, national consumption is 6 percent higher than under the existing situation (Scenario 1). At the same time, however, higher prices in major surplus-producing smallholder regions and commercial farming areas results in 7 percent greater supply. Lower marketing margins (from more direct transport routes and a shift to lower-cost informal millers) may therefore stimulate national maize supplies without adversely affecting consumer prices and household food security. Further analysis is required to examine the effects of decontrol on self-sufficiency and net trade under alternative weather scenarios.

CONCLUSIONS AND POLICY IMPLICATIONS: TOWARD THE DEVELOPMENT OF INFORMAL MARKETS

The preliminary results support the conclusion that movement restrictions on maize cause one of the most important income transfers in the organization of the grain-marketing system. This regulation restricts the supply of low-cost maize meal in urban areas, impedes private maize trade between surplus and deficit areas, and induces a circuitous, transport-intensive, and high-cost flow of national grain supplies. Furthermore, movement restrictions tend to force marketed grain surpluses in both commercial and smallholder areas into the GMB system, which consequently confines grain access mainly to the large industrial buyers. The combination of superfluous transport costs and relatively high milling margins of urban millers results in inflated prices for staple maize meal, especially in urban areas. This appears to be a major cause of food insecurity and loss of real income among grain purchasers.

The model indicates that relaxation of movement controls would result in substantial trade between surplus smallholder areas and urban centers. Smallholders close to major urban centers would receive higher prices, while urban consumers with a preference for informally milled meal would pay lower food prices. Results also indicate that movement decontrol would favorably affect the GMB's domestic trading account and increase national exportable surplus. Relatively little trade takes place between surplus and deficit rural areas. The existing weak road networks result in high transport costs between most smallholder areas.

The major difference between movement decontrol of smallholder areas only and full movement decontrol is that commercial farmers capture most of the benefits from full decontrol. Grain sales and incomes in surplus smallholder areas are appreciably lower in the latter case. Urban consumers would pay lower prices for maize meal in both scenarios.

The empirical results also pertain to a hypothetical system of competitive informal markets with the capacity to meet dramatic increases in volumes. The results therefore refer to *potential* gains to market reform. The ability to exploit this potential will require complementary government support for new entry and investment in private grain distribution, storage, and milling. The nature and severity of other constraints to private investment not directly related to grain-marketing policy must be identified and addressed if market reform is to result in market development.

Two independent surveys of rural and urban grain and meal traders indicate that government policy restrictions, ambiguous regulations, and limited capital, transport, and storage are major barriers to entry (Chisvo et al. 1991; Kinsey 1991). As an example of the latter, only 60 percent of the rural traders surveyed by Chisvo et al., owned a vehicle. Less than 50 percent owned a vehicle with the capacity to carry more than 20 bags of grain. Shortage of working capital also hinders investment in vehicles and economies of scale in distribution. Inadequate foreign-exchange allocations for engines has created long backorders of small-scale milling equipment.

A major implication of the foregoing is that the existing dominance of trade in maize meal rather than grain—while resulting in artificially high prices for consumers—overcomes constraints faced by traders.

Many traders have found commercial maize-meal trading to be a convenient substitute for grain trading because (1) most commercial millers deliver their meal directly to traders' shops, even in rural areas, permitting them to earn a 9 percent markup (set by government) for simply stocking a product that is delivered to their doorstep; (2) many traders buy commercial meal on credit; (3) commercial meal is delivered monthly, relieving the trader of the risks and costs of storage; (4) the trader avoids the information and transaction costs of having to locate buyers within surplus areas and performing bulking functions that would be necessary with grain trading; and (5) the demand for commercial meal is guaranteed by controls on grain movement and by the extraction of grain out of rural areas by the GMB. The commercial maize-meal distribution system thus eliminates critical transport, credit, storage, and informational constraints that grain trading would present, entrenching incentives for traders to deal in commercial meal rather than grain. However, the system creates high costs to consumers.

The majority of traders engaged in assembly and wholesaling appear to be uncertain of the legality of grain trading (Chisvo et al. 1991). While the ambiguity of trading regulations has not precluded the development of informal trade, such trade has been of lower volume and higher cost than would be the case if the rules were clear and government took steps to actively support intrarural trading activity.

These points suggest that policy reform, while necessary, is insufficient to induce the desired response by the private sector. Increased investment and new entry to develop rural grain markets requires active government support to relieve the transport, storage, credit, and informational constraints associated with grain trading. Such government support could include

- Allocating foreign exchange for importation of small-scale milling equipment;
- Promoting local metal-manufacturing industries that produce parts needed by small mills;
- Removing import restrictions and bureaucratic impediments associated with importing productive equipment and vehicles;
- Ensuring that grain is available for purchase at GMB depots by all individuals or businesses in any quantity above the current minimum of one bag; and
- Allowing anyone to become a legitimate grain buyer or seller instead of requiring licenses and prerequisites that restrict entry into grain trading.

These public investments and policy changes would be consistent with the government's current initiatives to promote emergent small-scale businesses under the Indigenous Business Development program. Once such milling and trading networks were in place to compete alongside the industrial milling sector, the costly subsidies on roller meal and superrefined meal could be removed, since low-income consumers would have access to lower-cost meal through informal channels.

APPENDIX 1: MARKET ORGANIZATION

UNIFORM PRICING AND MOVEMENT CONTROLS

To ensure a consistent flow of maize meal to urban consumers, the government of Zimbabwe has influenced prices and distribution through a highly controlled and centralized maize-marketing system. The official grain-marketing system features a predominantly one-way flow of grain from rural to urban areas and is characterized by centralized urban milling and storage facilities (Figure 9.1). Maize may be sold through the official system to one of three procurement arms of the GMB: GMB depots, normally located in urban centers or growth points; GMB collection points located in rural smallholder areas; and Approved Buyers, that is, licensed private traders who buy on behalf of the GMB. Private maize trading within smallholder areas was never banned, but is nevertheless circumscribed by numerous government regulations:

- Smallholder maize, unless destined for a GMB depot, is prohibited from moving across the boundaries of urban or commercial farming enclaves.[4] Since these areas contain virtually all of the country's main roads, this regulation effectively blocks private grain trade between noncontiguous smallholder areas or from smallholder areas to urban consumption centers.
- Maize may not be moved privately from commercial farming areas to smallholder areas.
- Once grain is sold to GMB collection points or approved buyers in smallholder areas, direct resale to consumers is prohibited. Instead, the grain must be forwarded to GMB depots, often a considerable distance from deficit rural areas. This effectively siphons supplies out of rural areas, tightens local supply-demand conditions and exerts upward pressure on rural market prices.
- The margin between the GMB purchase price and selling price to urban millers since 1986/87 has been only half of the GMB's actual operating costs (AMA 1990). The combination of consumer price subsidies and restrictions on direct trade between surplus and

---

[4] Grain-marketing policy divides the country into "Zone A" areas, which include all urban and European commercial farming areas, and "Zone B" African smallholder areas. The Zone B areas are geographically scattered and are essentially enclaves within Zone A. Even between contiguous smallholder areas, the scope for trade is limited by poor or nonexistent road infrastructure.

Figure 9.1—Formal and informal maize distribution channels

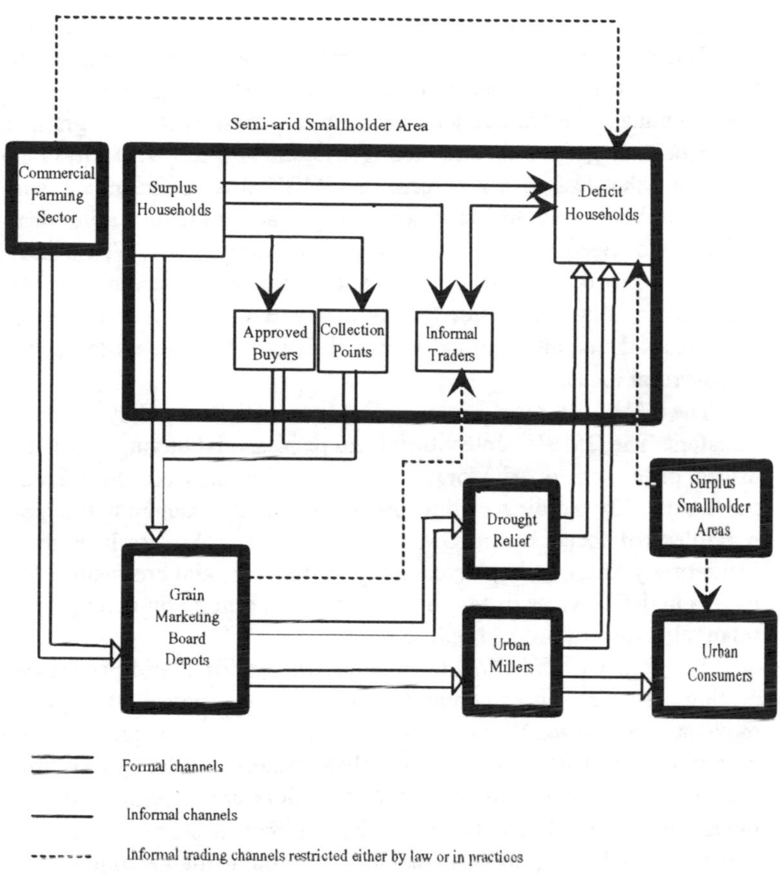

deficit rural areas has encouraged the consumption of urban-milled meal in deficit rural areas. Panterritorial prices for commercial-milled meal further extend the dominance of the official distribution system even in the most distant regions. These direct and indirect subsidies in the official marketing channel substantially narrow the scope for intrarural private trade.

- The rules governing resale of GMB grain to rural traders are subject to a variety of interpretations. In theory, any individual may purchase grain from GMB depots within smallholder areas (Grain Marketing Act, 1966, Section 21). Yet 13 of 15 GMB depot managers interviewed stated that they do not permit sales to informal traders suspected of reselling the grain, due to perceptions that they would exploit poor households needing grain (Chisvo et al. 1991).

This set of regulations governing movement and resale means that the choice of market channel at the farmer-first handler level largely predetermines the subsequent flow and accessibility of grain at subsequent stages in the marketing system (Figure 9.1). It is not surprising that less than 2 percent of GMB's total maize intake since 1980 has been sold to consumers or private traders. Large urban millers, stockfeeders, and brewers have accounted for 77, 8, and 6 percent of GMB sales since 1980. The remaining 7 percent has been used for food-aid purposes. Stocks at GMB depots in town centers throughout the country normally have little effect on access to grain in distant rural areas.

The GMB's uniform pricing policy is essentially a policy of income transfers. The GMB's panterritorial and panseasonal buying and selling prices offer subsidized storage and transport services to selective purchasers. By holding selling prices constant throughout the year, regardless of location, the GMB cross-subsidizes buyers later in the marketing year by taxing buyers early in the year, and cross-subsidizes buyers in deficit areas distant from production centers by taxing buyers relatively close to production centers.[5]

The ability of the GMB to practice panterritorial and panseasonal pricing requires corresponding policies that control the private movement of maize. Without such controls, maize producers in areas involving low marketing costs to urban centers would almost surely contract directly with the large urban millers and stockfeeders, thus bypassing the GMB. The GMB would lose market share on the routes that it currently uses to cross-subsidize its unprofitable routes in the more distant, semi-arid communal areas. This kind of decontrol would

---

[5] Since the GMB incurs progressively higher storage costs as the year progresses—keeping its selling price constant throughout the year—those who buy late in the season do not pay a price that reflects these storage costs, while those who buy early in the season help defray the storage costs of the late-season buyers. Similarly, since it is more expensive for the GMB to transport grain to Bulawayo than to Harare, given the geographical concentration of maize surpluses in the Mashonaland provinces the panterritorial selling price of the GMB essentially makes the Harare maize consumers help defray the transport costs of transporting grain to the Bulawayo consumers.

therefore exacerbate GMB operating losses rather than reduce them.[6] Therefore, the merits of regional versus spatial pricing for maize cannot be considered apart from movement decontrol.

MOVEMENT CONTROLS, THE MILLING SECTOR, AND URBAN FOOD SECURITY

Movement decontrol would also present serious implications for the maize milling industry. Currently, urban maize milling is dominated by four large private firms: National Foods, Blue Ribbon Foods, Midlands Milling Company, and Triangle Milling Company. National Foods alone handles about 65 percent of the market, while National Foods and Blue Ribbon combined handle 85 percent. These millers produce two types of maize meal: superrefined meal (60 percent extraction rate) and roller meal (85 percent extraction rate). Millers currently buy maize from the GMB and sell to retailers at government-controlled prices. Maize-milling margins are based on cost of production data supplied by millers.

Informal maize millers, by contrast, are restricted from procuring grain to mill in urban areas, because the GMB has in practice reserved its grain for the large industrial buyers, and because movement controls prevent informal traders from legally transporting grain into urban areas. As a result, the government has conferred a de facto monopoly on industrial millers, even though their margins are two to three times higher than those of small-scale millers.[7]

---

[6] While this chapter does not explicitly deal with panseasonal pricing, it should be noted that the relaxation of maize movement without modification of GMB panseasonal pricing would create incentives for producers and millers to contract directly for early-season deliveries, before storage costs bid the wholesale market price above the GMB selling price. Later in the season, industrial buyers would switch and attempt to buy from GMB at its uniform selling price. With the loss of sales early in the year, when storage costs are low, the GMB would no longer be able to cross-subsidize buyers later in the year. In this environment, it is doubtful that GMB could, without allowing spatial and temporal differentiation in its pricing, continue to perform the politically crucial functions of national security stockpiling and price stabilization. Nor is it clear that these tasks could be immediately assumed by the private sector.

[7] Informal small-scale milling margins were established from household surveys during 1990/91 together with before-and-after weight measurements of maize processed through a sample of hammer mills in Buhera and Mberengwa communal lands (Chisvo et al. 1991).

Lacking any threat of competition from informal millers, the industrial millers, whether by choice or circumstance, are able to operate a higher-cost system without losing market share (Figure 9.2). While available data indicate that maize meal could be sold through informal channels for 10-15 percent less than the control price of industrial roller meal, this option is blocked by policy to the majority of low-income consumers. Government regulation and pricing policy therefore appear to create incentives that perpetuate the distribution of more expensive meal, catering to higher-income tastes, with potentially adverse consequences for nutrition and incomes among the urban poor. The effective demand for the informally milled meals in urban areas is not well established, because grain-market regulations have historically blocked the informal sector from moving grain into urban areas and undercutting the prices of meal offered through the GMB/industrial milling system. However, evidence of demand for maize meal from informal mills is indicated by the following:

- Results of a recent household survey in four peri-urban areas of Harare indicate that there would be a moderate demand for straight-run meal among lower-income groups, after taking into account its cost discount. Sixty-two percent of the low-income group stated that they would purchase straight-run meal if it were 12 percent cheaper than roller meal and available in convenient bag sizes (Jayne et al. 1991). Relatively few of the high- and medium-income groups stated an interest in straight-run meal, even at substantial price discounts to the more refined meals. These findings indicate a potential for self-targeting, that is, a subsidy on straight-run meal would be conferred mainly to the poor.
- These surveys indicate that a portion of urban consumers are already consuming straight-run meal obtained from urban hammer mills. Maize produced around Harare as well as maize transported illegally from rural areas is milled by urban hammer millers into straight-run meal for urban consumption. Rapid appraisal surveys by the University of Zimbabwe/Michigan State University indicate that there are about 30-40 mills operating in the greater Harare area. These millers indicate that they operate at nearly full capacity during the several months after harvest, until local production is exhausted, after which throughput drops dramatically.
- An unknown quantity of maize is purchased illegally from commercial farms and subsequently milled into straight-run meal for consumption by households in nearby peri-urban areas.
- Straight-run meal accounted for approximately 5-8 percent of industrially milled maize meal before its manufacture in convenient

Figure 9.2—Price structure in the official grain marketing system and potential changes in response to market reform

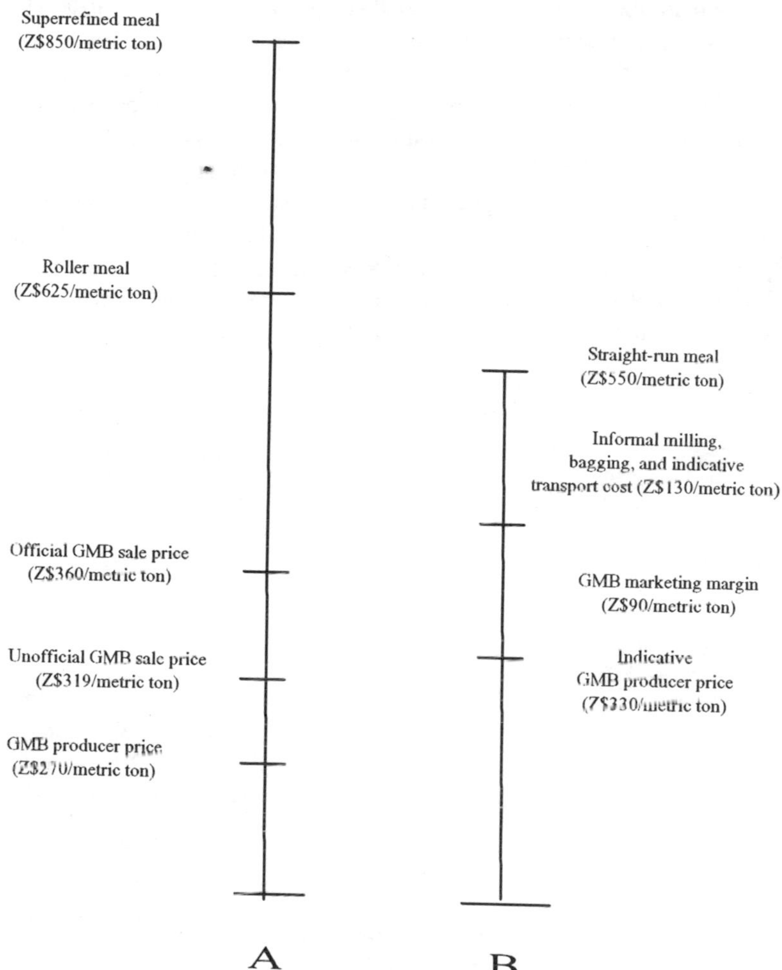

Notes: Column A is the official market price structure (Z$/metric ton, 1991/92 marketing year), and Column B is indicative price levels with legalization of informal milling sector in urban areas. GMB is Grain Marketing Board.

bag sizes was discontinued in 1979 (C. Robinson, pers. comm. 1991). While data is not available on consumption by income category, it is believed that this demand was concentrated primarily among the poor and in the southern portions of the country. The anticipated decline in real wages over the next several years may increase the demand for less-expensive staple food products.

This discussion highlights the potential difficulties associated with swift and wholesale liberalization of the marketing system. While the government of Zimbabwe is aware of the need for substantial changes to the organization of the grain-marketing system, it may be viewed as somewhat reckless to proceed without adequate analysis to guide the process of reform and to anticipate how such changes will affect price levels, supplies, food consumption, and income levels of various socioeconomic and regional groups in the country.

APPENDIX 2: MODELING TRADE FLOWS IN A DUAL MARKETING SYSTEM: THE CASE OF ZIMBABWE

ANALYTICAL FRAMEWORK

The analytical framework presented below is based roughly on Cole and Park (1983) and Roemer (1986).

Smallholders have two channels to which to sell grain: the GMB and informal buyers. The right-hand panel of Figure 9.3 represents the formal (GMB) market, while the left-hand panel represents the informal market. The supply curve (S) is net of own subsistence production. Survey evidence indicates that the proportion of total marketed maize surpluses sold to the GMB is a function of the supply-demand conditions in a particular region: the more surplus the area, the higher the proportion of grain sold to the GMB. Yet even in grain-deficit areas where informal maize prices are substantially above the GMB producer price, a corner solution rarely occurs, that is, neither the GMB nor informal buyers capture all of the marketed surpluses. The most important reasons why farmers sell to the GMB despite higher informal prices are (1) cash needs after harvest induce deliveries to the GMB at

Figure 9.3—Formal and informal market interactions: a surplus area

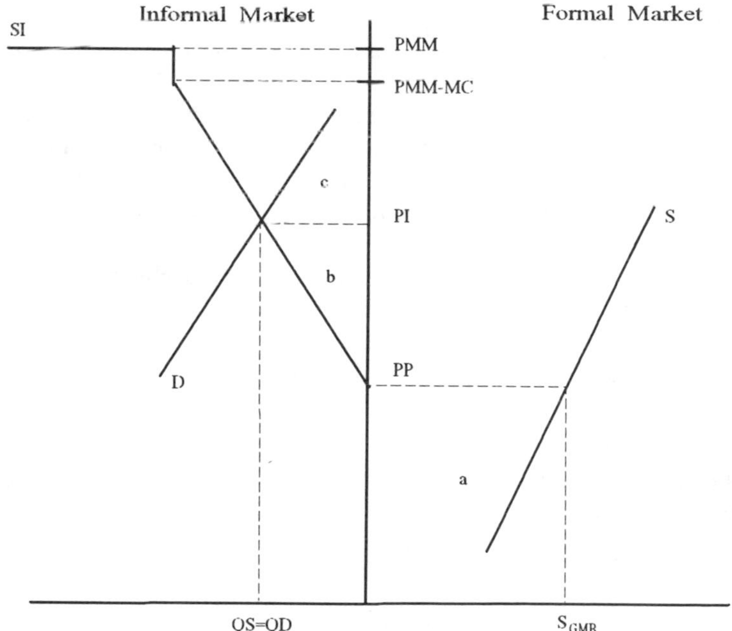

its panseasonal producer price (many informal buyers are not operating until later in the season); (2) repayment of credit from the Agricultural Finance Corporation is tied to crop sales to the GMB; and (3) after adjusting for transactions costs, GMB producer prices may be more attractive than informal prices for farmers near a depot, collection point, or approved buyer.

In Figure 9.3, the GMB producer price (PP) and stochastic weather determine the quantity of grain sold to the GMB (S_{GMB}) as well as the intercept of the residual informal supply curve (SI). It is assumed that the slope of the supply curve is the same for the informal and formal markets.

Similarly, grain-deficit smallholders have two channels from which to buy grain: industrial maize meal from retail shops and informal trade. In severely grain-deficit areas, the price of maize grain tends to approach the controlled price of industrial meal (PMM) minus informal milling costs (MC). Because industrial meal is readily available even in remote rural areas, PMM - MC is effectively a ceiling price for exchange within the informal market.[8] At this price, the supply of maize accessible to smallholders is perfectly elastic.

The intersection of the informal supply and demand curves yields the informal market-clearing price (PI) and quantity exchanged (QS = QD). Figure 9.3 illustrates a relatively maize-surplus region. Producer surplus is a + b, while consumer surplus is c.

Figure 9.4 illustrates a deficit area, in which the demand curve intersects the supply curve at its horizontal portion. In this case, demand outstrips the local supply of grain, and the less-preferred industrial maize meal must be purchased to meet residual demand. QS = QD represents the quantity of grain exchanged locally, and QMM - QD is the quantity purchased of industrial maize meal. Restrictions on private grain movement between regions prevent grain from flowing into such areas from surplus regions. In spite of local grain shortages, some maize (S_{GMB}) was funneled out of the region through the official marketing channel.

This framework is consistent with the observed phenomenon of grain flowing out of many smallholder areas through the state marketing channel while relatively expensive urban-milled maize meal flows into these areas (Table 9.3). Of the grain-deficit households in these

[8] Price-monitoring surveys indicate that the actual price of industrial meal is seldom more than 5 percent above the control price even in the most remote areas (Chisvo et al. 1991).

Figure 9.4—Formal and informal market interactions: a deficit area

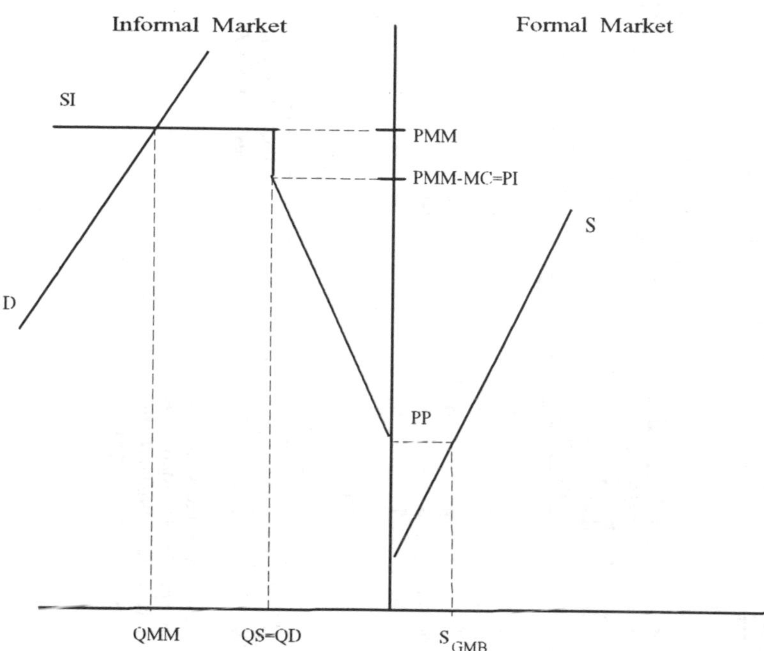

areas, 75 percent stated that they bought urban-milled meal simply because grain was not available locally. Therefore, the sale of grain "surpluses" to the GMB, while giving the illusion of self-sufficiency, has masked and even contributed to food insecurity in many smallholder areas (Jayne et al. 1990a).

Figure 9.5 shows the impact of raising the GMB producer price in certain deficit areas on informal prices under restrictions on maize movement. Some analysts have argued that higher GMB producer prices in the deficit areas will stimulate production there, thereby reducing the need for maize meal inflows, food aid, and the transport costs associated with these. However, an increase in the producer price from PP to PP' will shift supplies to the GMB that might otherwise have been sold informally. Once delivered to the GMB depots, the grain is inaccessible to most rural consumers. The informal market supply curve consequently shifts to SI', raising the informal market price and inducing greater inflows of expensive industrial maize meal. It is clear that consumer surplus is reduced in this scenario without an appreciable increase in producer surplus.

Table 9.3—Aspects of household grain-marketing behavior in selected smallholder farming areas

Communal Area	Share of Households That Are Net Grain Purchasers	Share of Total Grain Sales by the 10 Percent of Farm Households Selling the Most Grain	Average Household Net Grain Sales: Surplus Households	Average Household Net Grain Sales: Deficit Households	Share of Total Grain Sold To: Grain Marketing Board or Licensed Agents	Share of Total Grain Sold To: Neighboring Households	Share of Total Grain Sold To: Private Traders	Share of Total Grain Purchased from: Grain Marketing Board	Share of Total Grain Purchased from: Neighboring Households	Share of Total Grain Purchased from: Private Traders	Share of Total Urban-Milled Meal Purchased from Shopkeepers
	(percent)		(kilograms)					(percent)			
High rainfall											
Gokwe (south)[a]	12	51	3,707	-183	86	8	6	7	80	13	0
Buhera (north)[a]	26	50	1,023	-252	69	16	15	16	70	1	13
Low rainfall											
Gokwe (north)[a]	59	59	1,118	-438	5	95	0	10	44	36	10
Buhera (south)[a]	57	72	973	-392	68	31	1	0	40	16	44
Runde[a]	61	74	1,465	-344	30	70	0	0	23	37	40
Mberengwa[a]	85	60	834	-483	43	57	0	26	15	17	42
Nata[b]	94	57	21	-301	0	100	0	0	7[c]		92
Ramakwebana[b]	96	68	340	-383	0	100	0	0	13[c]		87
Semukwe[b]	98	62	46	-352	0	100	0	0	21[c]		79

Sources: University of Zimbabwe/Michigan State University/ICRISAT, *Grain Marketing Surveys* (Harare: University of Zimbabwe, 1990); B. Hedden-Dunkhorst, "The Role of Small Grains in Semi-Arid Smallholder Farming Systems in Zimbabwe: Preliminary Findings" (SADCC/ICRISAT, Matapos, Zimbabwe, 1990, mimeographed).

Note: Rainfall was average to moderately below average during the relevant production years in all survey areas.

[a] The results of these UZ/MSU/ICRISAT surveys pertain to the period from April 1989 to March 1990.

[b] The results of these UZ/MSU/ICRISAT surveys pertain to the period from November 1988 to October 1989.

[c] The distinction between purchases from households and private traders was not made in this study.

Figure 9.5—Impact of raising the Grain Marketing Board producer price in a dual market system

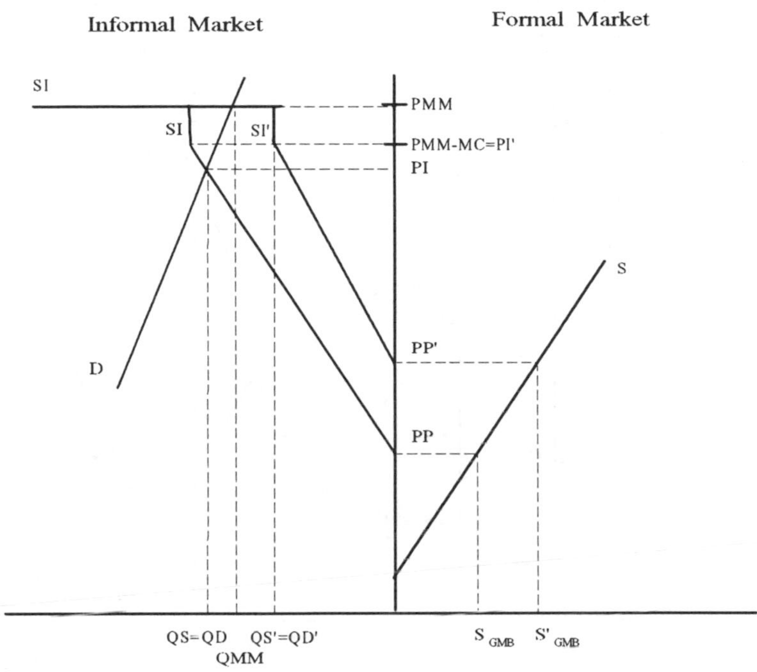

Figure 9.6 shows the impact of removing movement controls between a surplus and deficit area, thus allowing direct trade to take place between them. The left-hand panel represents a deficit area, the right-hand panel represents a surplus area, and the middle panel represents excess supply and demand curves (ES and ED) originating from these regions, following standard trade theory. The intersection of these excess supply and demand curves (after accounting for a marketing cost wedge) determines a new market-clearing price in the informal markets. Informal prices in the deficit region fall from PI_d to PI_d', while prices in the surplus region rise from PI_s to PI_s'. The volume of informal interregional trade is QD – QS in the deficit region (which is equal to QS – QD in the surplus region). Note that in this case, trade between the rural areas reduces the inflow and consumption of industrial milled meal in the deficit region. There is a net efficiency gain (consumer + producer surplus) in both the surplus and deficit areas.

Full decontrol of maize movement and prices would result in the elimination of market segmentation. Prices would then be determined

Figure 9.6—Effect of grain movement decontrol on surplus and deficit areas

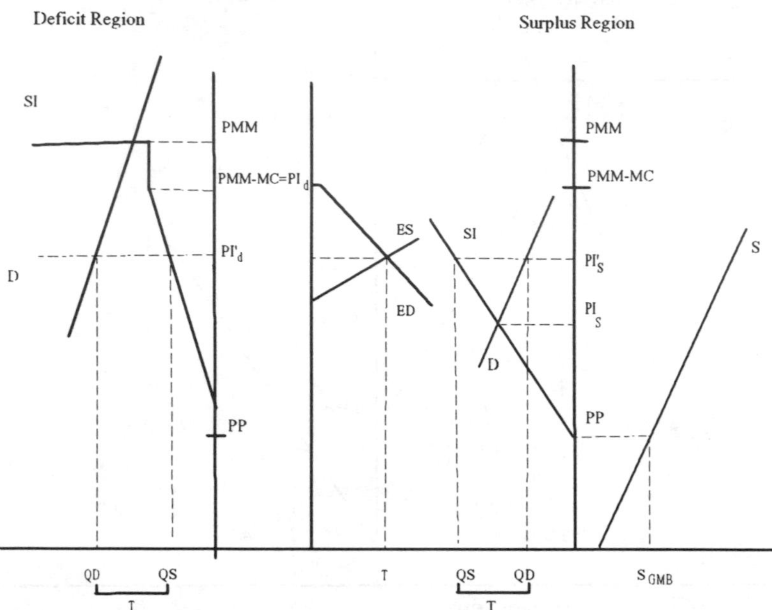

by supply and demand conditions. The GMB could operate viably in such a system only if it bought and sold grain according to these regional supply and demand conditions. The resulting abolition of panterritorial prices would in turn prevent millers from offering a panterritorial maize-meal price, hence the price ceiling on maize trade would be removed.

This analytical model operates as follows:

1. Maize supply functions are estimated econometrically for each producing region (13 smallholder areas and 5 commercial areas). Econometric specifications and the procedure for estimating maize demand are described below.

2. The GMB producer price, chosen exogenously, determines the portion of total surplus production sold to the GMB and the residual informal supply curve. The price of industrial maize meal is the price ceiling in the informal market; at this price, supply is completely elastic.

3. The informal price, which is derived from supply and demand conditions in the smallholder area in question, is nevertheless influ-

enced by government pricing decisions in the official market.

4. The quantity of maize demanded and supplied in the official market, aggregated across all regions, equals national net maize surplus (which is either traded or stored).

5. The model may be run for a variety of rainfall conditions. This would have the effect of changing the intercept of the supply curve in the official market and, under some conditions, in the informal market.

6. Under the scenario of controls on grain movement, each region is in autarky, except for the movement of industrially milled meal to meet demand in deficit regions. Under the scenarios involving decontrol of maize movement, the model is similar to standard trade models where excess supply and demand curves are determined from the supply and demand curves in the respective regions. Excess supply and demand determine a unique informal price that is modified by relevant transport and processing costs. Sensitivity analysis on these margins is possible to examine the robustness of trade flows and prices to various assumptions about the competitiveness and efficiency of the informal market.

7. By aggregating across regions, national supply to the GMB and sales of industrially milled meal can be derived. The residual, after adjusting for milling extraction rates and demand for GMB grain from stockfeeders and brewers (which is a relatively small part of the market and treated as a constant), is national surplus, that is, end stocks plus net exports.

DATA USED TO RUN THE MODEL

Maize supply

Supply elasticities used in the model are based on econometric estimation of structural equations for annual GMB intake over the 1978 to 1990 period. The supply functions are estimated using regionally disaggregated price, weather, and GMB maize intake data (AMA, various years; GMB files; National Meteorological Service, various years). Smallholder equations were of the form

$$S_{it} = a_0 + a_1(PP^*_t/CPI_{t-1}) + a_2(PPSUB_{t-1}/CPI_{t-1})$$
$$+ a_3(RAIN_{it}) + a_4(AFC_{it-1}) + e_{it}, \qquad (1)$$

while commercial sector equations were of the form

$$S_{it} = b_0 + b_1(PP^*_t/PF_{t-1}) + b_2(PPSUB_{t-1}/PF_{t-1}) + b_3(RAIN_{it}) + b_4(TREND_t) + v_{it}, \quad (2)$$

where

S_{it} = sales to GMB by regional group i in harvest year t;

PP^*_t/CPI_{t-1} = expected GMB producer price to be announced in marketing year t/t+1 deflated by the consumer price index at time of planting;

$PPSUB_{t-1}/CPI_{t-1}$ = producer price of a major substitute cash crop in region i at the time of planting deflated by the consumer price index at time of planting;

$RAIN_{it}$ = rainfall condition variable;

AFC_{it-1} = number of smallholders receiving credit from the Agricultural Finance Corporation in region i during the planting year;

PP^*_t/PF_{t-1} = expected GMB producer price to be announced in marketing year t/t+1 deflated by the price of nitrogen fertilizer at time of planting;

e = error term;

$PPSUB_{t-1}/PF_{t-1})$ = producer price of a major substitute cash crop in region i at time of planting deflated by the price of nitrogen fertilizer at time of planting. The substitute crops chosen were tobacco in the commercial equation, cotton in the Mashonaland and Midlands/ Manicaland smallholder equations, and sunflower in the remaining communal equations;

$TREND_t$ = a time trend to capture the effects of excluded time-correlated factors; and

v = error term.

Since GMB producer prices for maize have been announced after planting time for the past decade, the maize supply equations must be formulated on the basis of price expectations, using information available to the farmer at planting to predict the likely price announced after harvest. It is well known that the government-determined maize producer price is influenced by the level of GMB maize stocks from the previous harvest and by recent price trends (Wright and Takavarasha 1989). This suggests a simple maize price expectations model of the form

$$PP_t^* = b_0 + b_1(ENDSTOCKS_{t-1}) + b_2(PP_{t-1}), \qquad (3)$$

where PP_t^* is the expected price to be announced by the GMB at harvest year t, $ENDSTOCKS_{t-1}$ are the GMB maize stock levels at the end of the previous marketing year, and PP_{t-1} is the GMB price announced in the previous year.

The estimated maize price elasticities of supply were as follows:

Mashonaland commercial: +1.62**
Other commercial: +0.28
Mashonaland smallholder: +0.76*
Midlands/Manica smallholder: +0.27
Masvingo/Matabele smallholder: +0.48
* significant at the 95 percent level; ** significant at the 99 percent level.

Maize Demand

Information on maize demand by region is not available in Zimbabwe. However, reasonable first-order approximations of maize-grain and maize-meal purchases may be obtained through a combination of household survey data across various regions and years. This provides a point on the demand curve for each region. After this, a price elasticity of demand of −0.4 and −0.8 was assumed for urban areas and rural areas, respectively.[9]

GMB Operating Cost Estimation

Estimates of the GMB trading account are modeled as a function of price, intake, demand, stock, and trade outcomes from the simulation model. The GMB trading account equation is based on Buccola and Sukume (1988) but adapted to account for regional variations in the GMB's maize-procurement costs. Region-specific transport and

[9] The elasticity of −0.8 for rural areas may be too high considering the low substitutability of foodgrains.

handling costs for the 1990/91 marketing year were provided by GMB personnel. The equation is

$$ER \cdot [fob \cdot NX \cdot D - cif \cdot NX \cdot (1 - D)] - STK - ADM + P2 \cdot DD - (P1 + t_i)S_i, \quad (4)$$

where

- ER = official exchange rate (\$US/Z\$);
- fob = GMB maize export revenue (Z\$/metric ton);
- NX = net maize exports (metric tons);
- D = dummy variable ($D = 1$ if $NX > 0$, $D = 0$ if $NX < 0$);
- cif = GMB maize import expenditure (Z\$/metric ton);
- STK = GMB stockholding costs [Z\$40 • ($endstocks_t$ – $endstocks_{t-1}$)/2];
- ADM = administrative costs (a constant, Z\$15/metric ton);
- P2 = GMB selling price (Z\$/metric ton);
- DD = demand for GMB maize (metric tons), calculated as the aggregation across regions of purchases of industrial maize meal, adjusted by the average grain-to-meal outturn rate, plus demand for maize by stockfeeders and brewers and others (treated as a constant in this analysis);
- P1 = GMB producer price (Z\$/metric ton);
- t_i = GMB transport and handling costs from region i to nearest demand center (Z\$/mt); and
- S_i = GMB maize intake from region i (Z\$/mt).

REFERENCES

AGRITEX (Agricultural, Technical, and Extension Services). Various years. Extension worker forecast estimates. Ministry of Lands, Agriculture and Rural Resettlement, Harare. Mimeo.

AMA (Agricultural Marketing Authority). Various years. *GMB report and accounts*. Harare: AMA.

Buccola, S., and C. Sukume. 1988. Optimal grain pricing and storage policy in controlled agricultural economies: Application to Zimbabwe. *World Development* 16 (3): 361-71.

Chisvo, M., T. S. Jayne, J. Tefft, M. Weber, and J. Shaffer. 1991. Traders' perceptions of constraints on informal grain marketing in Zimbabwe: Implications for household food security and needed research. In *Market reforms, research policy and food security in the SADCC region*, ed. M. Rukuni and J. B. Wyckoff, 120-138. Proceedings of the Sixth Annual Conference on Food Security Research in Southern Africa, 12-14 November 1990, University of Zimbabwe/Michigan State University Food Security Research Project, Department of Agricultural Economics and Extension, Harare.

Cole, D., and Y. C. Park. 1983. *Financial development in Korea, 1945-78*. Cambridge, Mass., U.S.A.: Council on East Asian Studies.

Grain Marketing Board. 1966. The Grain Marketing Act (CAP 113), 1966. In *Grain Marketing Board Circular*. Harare: GMB.

_____. 1991. Response from the management of the Grain Marketing Board to the Economic Structural Adjustment Program. Harare: GMB, Planning Unit.

Hedden-Dunkhorst, B. 1990. The role of small grains in semi-arid smallholder farming systems in Zimbabwe: Preliminary findings. SADCC/ICRISAT, Matopos, Zimbabwe. Mimeo.

Jayne, T. S., and M. Chisvo. 1991. Unravelling Zimbabwe's food insecurity paradox: Implications for grain market reform in southern Africa. *Food Policy* 16 (August): 319-329.

Jayne, T. S., M. Chisvo, S. Chigume, and C. Chopak. 1990a. Grain market reliability, access and growth in low-potential areas of Zimbabwe: Implications for national and regional supply coordination in the SADCC region. In *Food security in the SADCC region*, ed. M. Rukuni, G. Mudimu, and T. S. Jayne. Proceedings of the Fifth Annual Conference on Food Security Research in Southern Africa, 16-18 October 1989, University of Zimbabwe/Michigan State University Food Security Research Project, Department of Agricultural Economics and Extension, Harare.

Jayne, T. S., M. Chisvo, B. Hedden-Dunkhorst, and S. Chigume. 1990b. Unravelling Zimbabwe's food insecurity paradox: Implications for grain market reform. In *Integrating food and nutrition policy in Zimbabwe*, ed. T. S. Jayne, M. Rukuni, and J. B. Wyckoff. Proceedings of the First Annual Consultative Workshop on Food and Nutrition Policy. University of Zimbabwe/Michigan State University Food Security Research Project, Department of Agricultural Economics and Extension, Harare.

Jayne, T. S., M. Rukuni, M. Hajek, G. Sithole, and G. Mudimu. 1991. Structural adjustment and food security in Zimbabwe: Strategies to maintain access to maize by low-income groups during maize marketing restructuring. In *Toward an integrated national food policy statement*, ed. J. Wyckoff and M. Rukuni, 8-50. Proceedings of the Second National Consultative Workshop, 10-12 June, University of Zimbabwe, Department of Agricultural Economics and Extension, Harare.

Kinsey, B. F. 1991. Private traders, government policies, food security and market performance in Zimbabwe. International Food Policy Research Institute, Washington, D.C. Mimeo.

National Meteorological Service. Various years. *Meteorological Bulletin*. Harare, Zimbabwe.

Roemer, M. 1986. Simple analytics of segmented markets: What case for liberalization? *World Development* 14 (3): 429-39.

University of Zimbabwe/Michigan State University/ICRISAT. 1990. *Grain Marketing Surveys*. Harare: University of Zimbabwe.

USDA (U.S. Department of Agriculture). 1988. *World food needs and assessment.* Washington, D.C.

Wright, N., and T. Takavarasha. 1989. *The evolution of agricultural pricing policy in Zimbabwe.* Working Paper 11-89, Department of Agricultural Economics and Extension. Harare: University of Zimbabwe.

Zimbabwe. 1982. *Transitional National Development Plan 1982-3 to 1984-5*, vol. 1. Harare: Government Printer.

_____. 1988. *First Five-Year National Development Plan, 1986-90*, vol. 2. Harare: Government Printer.

_____. 1991. *A framework for economic reform (1991-95).* Harare: Government Printer.

10
Grain Market Liberalization Should Not Ignore Legitimate Social Goals: The Case of the Grain Marketing Board in Zimbabwe

Share J. Jiriyengwa

Zimbabwe has a single-channel marketing arrangement for its controlled agricultural products. The Grain Marketing Board (GMB) handles six controlled crops and five regulated crops.[1] The controlled products are maize (both white and yellow), soybeans, groundnuts, white sorghum, sunflower seeds, and coffee. The regulated products are red sorghum, finger millet (*rapoko*), pearl millet (*mhunga*), edible beans, and rice.

The government announced its commitment to reform the agricultural marketing system in the budget statement of July 1990. Pressure for reform came mainly from the budget deficit and negative per capita growth in agriculture. The average agricultural growth rate was estimated at 2.2 percent a year between 1980 and 1988 (World Bank 1991), compared with an estimated population growth of 3.0 percent a year. The decline in real producer prices over recent years added to the pressure from farmers for reform. The increasing parastatal losses and their adverse impact on the national budget and pressure from international agencies to reduce subsidies have all contributed to the call for reform.

The documented reforms have dwelt mainly on the parastatal losses, specifying targeted reductions in those losses over a four-year period. The reforms were primarily concerned with reducing budgetary costs and were termed "deficit-reducing measures" by the parastatals concerned. Rather than view the reforms as a strategy to rejuvenate

---

[1] A product declared to be "controlled" by Zimbabwe's Minister of Lands, Agriculture and Rural Resettlement cannot be marketed through any channel other than the parastatal organization designated for this purpose. The parastatal can, however, appoint an agent to purchase such produce in designated areas on condition that the agent is allowed to resell only to the parastatal. For a "regulated" product, free trade is allowed in all areas, but the parastatal (residual buyer) is obliged to purchase all surpluses offered to it by producers.

agricultural growth, deficit-cutting has been regarded as an end in itself.

Treating reforms as budgetary cost-saving programs has led to the adoption of a simplistic view that all that is required is the trimming of public enterprises and privatizing of their functions. Parastatals drew up cost-cutting measures and seized the opportunity to point out a number of financially nonviable operations.

This paper argues that the reforms should aim primarily at restoring vibrancy in the agricultural industry. It starts from the premises that

- The private sector in Zimbabwe is too weak to support sustained agricultural growth. Having suffered decades of official oppression, it requires more active support than the mere removal of physical controls, and the results can be felt only in the long term, much longer than the stipulated reform period of up to 1995.
- It is necessary to maintain a viable public sector to protect the gains made in the past (notably access to markets by small and remote producers) while a private sector evolves.
- The GMB is both a de facto and de jure monopoly. As such its abolishment will leave a gap.

MAIN FEATURES OF THE CURRENT MARKETING SYSTEM

For purposes of marketing controlled products, the country is divided into areas A (covering all commercial farms and urban/industrial settlements) and B (covering all communal lands). Controlled products cannot be moved from one area into another unless they are being sold to, or purchased from, the GMB. For ease of policing, the same products cannot be moved from one part of the country classified as B to another of the same classification where such movement has to transit a part of the country classified as A (trade between noncontiguous communal lands). This arrangement has been blamed for separating suppliers of controlled products from the demanders of such products and, hence, stifling effective private trade even in those areas where the law allows it.

Producer prices and selling prices of controlled agricultural products are "fixed" by the government on a formula that weights farm viability, implications for parastatal trading accounts, and consumer welfare and equity more than market forces of supply and demand. Producer prices are therefore set relative to farm costs of production on a cost-plus basis. They are panterritorial and panseasonal. Consumer prices are set relative to changes in purchasing power of industrial wages, with any differences between these and the GMB's procurement and marketing costs being met by state subventions. These too are panterritorial and panseasonal. These pricing formulas have taken Zimbabwe's prices out of line with world agricultural prices. The parastatal mode of operation has been accused of "breeding market-illiterate" producers and consumers and failing to synchronize supply and demand of controlled agricultural products. The pricing system is said to have complemented the movement restrictions to further suppress private grain trade (Jayne et al. 1991).

The GMB deals with supply-demand gaps through administrative allocations based on processors' established capacities and historical market shares. This not only denies the operation of market signals in directing resources to alternative uses but also protects established processors against competition from new entrants to the industry. Innovation among existing processors is inhibited, and the GMB is forced to store such commodities for them throughout the year, with adverse impacts on its trading accounts. On several occasions, the GMB has had to force expressers and millers to withdraw their outstanding allocations by the end of the year, casting doubts on whether the raw materials are in short supply relative to processors' capacities (GMB 1991a).

The GMB has operated as a tool for the implementation of government policy but has had minimal influence in policy formulation. The board's expertise in market dealings was fed to decisionmakers indirectly through consultative meetings with the Agricultural Marketing Authority (AMA), which had a policy advisory role to the government.

The GMB's investment mainly centered around the depot network, a network of marketing points that were heavily biased in favor of receiving marketable surpluses from producers and selling them to consumers in major consumption areas. Before 1980, this network was well established only in commercial farming and major consumption areas. The political and social objectives of the postindependence government were committed to redressing this imbalance, and the GMB was specifically tasked with the expansion of the depot network in communal areas. The major criterion used to establish this network was the reduction of marketing costs for communal farmers, and as transport to the market was the major cost item for these farmers, distance became the dominant factor in determining depot construction. The government instructed that depots be constructed in communal areas until every farmer (even in deficit areas) was "within 45 kilometers of a marketing outlet" (GMB 1991a). Thus sociopolitical considerations for the communal economy took precedence over its economic and financial viability. The board's investments were therefore not "cost sensitive" and were unresponsive to throughput variation.

The AMA raised all finance for the four agricultural boards; however, it also raised foreign finances for the government. It had extensive "off-shore" borrowings, and the repayments were obviously subject to currency-exchange fluctuations. Although the boards required no foreign currency, and the domestic money market was willing to lend money to the AMA, the boards found themselves in the unfortunate position of having to pay for the currency-depreciation costs. The benefits from such currency inflows to Zimbabwe as a whole have never been disputed, but the charging of "exchange losses" added a negative aspect to the boards' trading results and hence to their implied operational efficiencies as well as artificially inflating apparent agricultural subsidies. With the continuous depreciation of the Zimbabwean dollar (Z$) the exchange loss factor rose to such levels that the effective interest the GMB paid on AMA borrowings exceeded the market rate.[2] Lack of control over these key variables had compromised management's liability on the final trading outcomes of the parastatals. It was difficult to formulate performance parameters upon which the operational efficiency of the organization could be judged. This weakened the management accountability process and removed incentives to more cost-effective management practices.

Three objectives are implied in the Act that authenticates the GMB mandate: (1) market stabilization; (2) national food self-sufficiency, with a heavy bias in favor of cheap food for the industrial wage-earning groups; and (3) commercial activities of buying/selling of grains (trading).

In practical terms, however, these objectives are not clearly delineated, since the cost of achieving each objective cannot be calculated. There has been some "cross-subsidization" between objectives, which has militated against efficiency.[3] The objective of

[2] However, the GMB still uses the nominal interest rate charged by AMA in costing its stockholding function. This rate is lower than the market rate and, given the increase in the selling price of stocks that is inevitable the following year, underpricing of the stockholding function could act as an incentive to hold stocks rather than sell and removes the need on the part of the GMB to adopt aggressive marketing strategies.

[3] For example, the cost of carrying the GMB's huge stocks has been passed on as the cost of holding "strategic stocks" and so charged to the food-security account, when in fact, except for only two years after independence, the stockholdings were a result of the stabilization function. Such costs should have been passed on as the cost of achieving the stabilization objective and not the food-security objective.

bringing the peasantry into the cash economy has not been officially incorporated in the GMB objectives, and the parastatals' achievements in line with this goal are not quantifiable in revenue terms. Yet their financial costs are incorporated in the trading accounts. The resultant distortions and misinterpretations of the parastatals' trading accounts tend to demoralize management.

ADVANTAGES OF THE CURRENT SYSTEM

Following decades of systematic marginalization of the peasant population, most communal lands are in the remote, usually semi-arid regions of the country. Landholdings are generally small and degraded, offering a limited choice of cropping alternatives. Given the ability of the commercial sector and a few high-productivity communal areas near urban centers to satisfy most of the viable demand for grains and oilseeds, the prices receivable by the majority of small producers would be severely depressed in the absence of an extensive, subsidized depot network in these areas.

Critics of the panterritorial pricing system argue that it depresses producer prices in remote deficit areas and that movement controls discourage traders operating to reduce consumer prices in remote areas. The author concedes that because deficit and surplus areas are not contiguous, there would be little additional intra- or intercommunal land trade. The current system ensures stable producer and consumer prices and allows the authorities to influence the choice of production, and has in the past been used to promote the production of maize.

The GMB recognizes that it is unlikely to be an effective instrument for selling grains directly to widely dispersed, poor, rural households. The board should merely perform a wholesaling function from positions convenient from the transport point of view and work toward the development and strengthening of private retailers. Further, the board does not consider it prudent to expect the objective of food security in its broader sense to be accomplished through one instrument or one organization. Food security in the GMB is thus defined in its narrower sense of ensuring adequate food supply at the national level and meeting food production goals. Pursuing food security at the household level would entangle the parastatal in a web of conflicting action plans that may end up destroying production initiative and seriously threatening national availability of grains. Household food security encompasses a variety of aids to market efficiency that a single organization cannot efficiently provide, such as availability of rural credit, promotion of business entrepreneurship, and developed infrastructure (USDA 1991).

DISADVANTAGES OF THE CURRENT SYSTEM

One of the main targets of reform is to reduce fiscal dependence. A number of factors have played a major role in the GMB's trading losses. The price system failed to synchronize supply and demand for grains, resulting in costly stockpiles. In a crude regression model, Muchero (1986) found a significant positive relationship between average monthly stocks held and GMB losses on the maize trading account. A vast depot network that was dictated mainly by social and political considerations resulted in declining average depot throughput (Figure 10.1). The increased number of purchasing transactions in rural areas reduced average transaction size, thus increasing unit handling costs. In 1983/84 there was an average of 8 metric tons per receipt; this had dropped to 1.7 tons in 1990/91. An analysis of the GMB's maize trading account shows a significant proportion of the deficit could be explained by the share of small producers' deliveries in total GMB purchases of maize (Jiriyengwa 1989). The taxes imposed by AMA borrowings abroad have contributed significantly to the board deficits. Finally, the general operating inefficiency promoted by weak accountability processes and the bureaucratic value of the organization have also contributed to high marketing margins. The costs (in current prices) of marketing maize since 1981 are depicted in Figure 10.2.

Increased parastatal losses exerted a downward pressure on producer prices payable to farmers. The producer-price formula incorporated some considerations for parastatal trading results and stockholdings; thus the accumulation of stocks and the board's losses militated against substantial producer-price increases. This particularly taxed commercial maize farmers who were well placed to service domestic urban demand and whose land had alternative employment to maize production. This farming group called for the dismantling of controls in maize marketing and started a steady switch away from maize to nonregulated export crops; for example, horticulture and tobacco. Communal farmers campaigned for the maintenance of the status quo—their access to, and competitiveness on, major grain markets was severely restricted, mainly by distances and fewer cropping alternatives. Further, they have an inherent dislike for private-

Figure 10.1—Indices of volume handled, average employees, and productivity

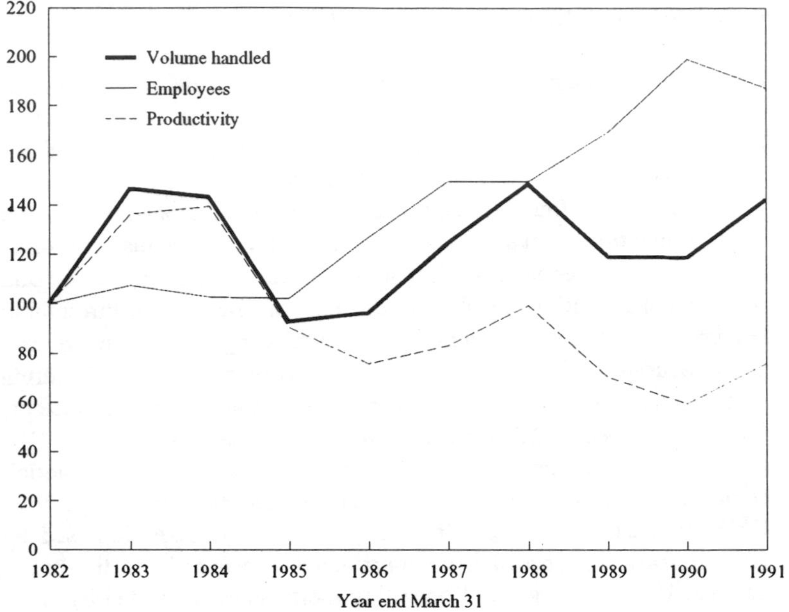

Source: Grain Marketing Board, "Proposals for Deficit Reduction," Background analysis to GMB Business Plan (GMB Planning Unit, Harare, 1991).

sector traders, whom they regard as exploiters, and all trade practices that deviate from the parastatal mode are regarded as exploitative.

In an attempt to contain the increasing trading losses, the authorities were forced to widen GMB selling-price margins over producer prices. This created an incentive to producers and consumers to bypass the parastatal, effectively creating a black market. For geographic and infrastructural reasons, communal farmers were not able to benefit from these illegal opportunities, which tended to involve the large milling companies.

The parastatal's mode of operation left no room for private trade in those areas where the parastatal was the designated residual buyer.

Figure 10.2—Finances of maize trading

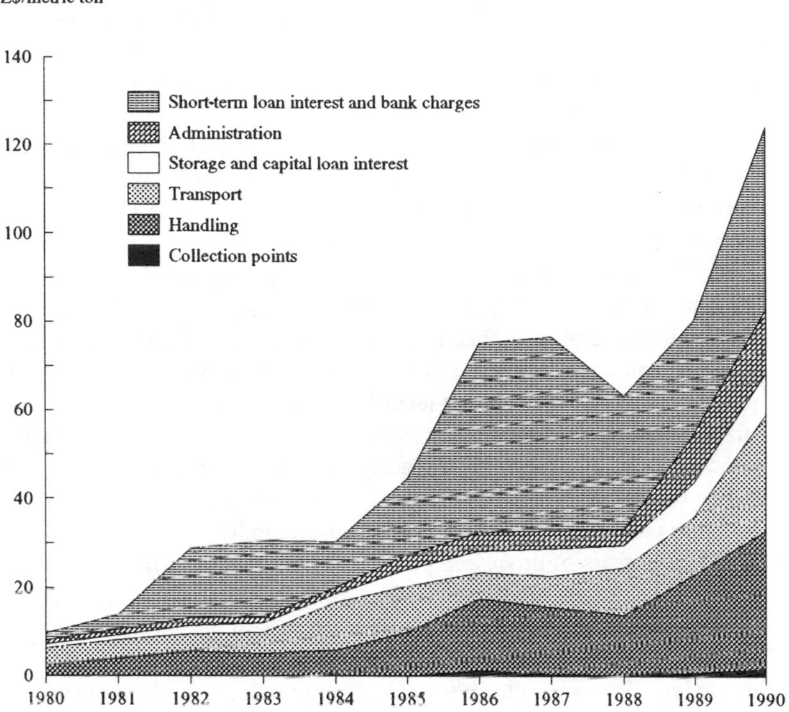

Source: Grain Marketing Board, *Annual Trading Accounts*, (Harare: GMB, various years).

Its selling margins, though reasonably wide in recent years, remained such that, given the distances to major markets and the transport bottlenecks in those areas, they left little room for reasonable profit-making by private traders. The licensed traders acting as GMB agents were tied in a contract that set the margins they would make. A serious drawback to private trade operation is the stipulation that resale of grains purchased can be only to the GMB.

The GMB operates on a principle of meeting domestic demand first, with exports as residual, for all crops except coffee. This prevented producers of exportables like soybeans and maize from benefiting from the upswings in world prices and, in recent years, from the continuous depreciation of the Zimbabwean dollar against major currencies. Such producers were heavily taxed and were denied opportunities to benefit from export incentives, including the recently announced export-retention scheme. As a result, there has been a notable shift away from politically sensitive crops such as maize and oilseeds toward tradables like tobacco.

The policy of panterritorial pricing also extends to processed grains and oilseeds products. With the policy that the GMB supplies processors in major consumption areas free of transport costs (Zone Center Policy), the processing capacity became concentrated in these urban areas. Panseasonal pricing has discouraged processors from developing storage facilities, and they have relied on the GMB's feeding raw materials to them on a "hand to mouth" basis. Administrative allocations for commodities considered to be in short supply discouraged new processors from establishing themselves in either remote places or large urban centers, adding an element of oligopoly to the geographic concentration of processing capacity. The transport and storage costs to the GMB contributed significantly to the parastatals' losses.

The concentration of processing capacity in large urban centers also taxed consumers of processed products in rural and small urban centers through increased transport costs. The oligopolistic nature of the processing industries made it easier for them to pass this double transport cost to consumers, and their margins have been regarded as excessive (Jayne et al. 1991). Yet above these wide margins the processors, especially millers, enjoy further subsidies that discount the nominal GMB selling price.

THE REFORM PROGRAM AND THE GMB

In the reform program announced by the government, the main issues relating to the GMB were

- Classification: the parastatal would be classified as an "entity with a valid social role and will be maintained in government hands."
- Subsidies: a targeted elimination of government subventions to the parastatal over a four-year period. "Whatever subsidies will remain will be small and transparent."
- Transparency: whenever the government requires the GMB to carry out a project or activity of a social or noncommercial nature, the government will provide financing for such activity through a direct budgetary subsidy or grant.

- Autonomy: a board of directors with greater independence will be put in place and charged with orientating the parastatal's operations toward commercial viability.
- Monitoring and accountability: this would be improved with the designation of a central unit for monitoring performance. A computerized monitoring system will be created with improved audit requirements.
- Competition: private trading in grains will be introduced in a phased manner until full competition is permissible by 1995. Movement restrictions in grain marketing will be phased out; pricing will be allowed to vary in line with market forces across regions and over time.

THE IMPLICATIONS OF REFORM AND PROPOSED OPTIONS

The reform program was welcome in principle to GMB management, but they are concerned with nonsynchronization of the deficit reduction program and reforms to policies that gave rise to the deficits (GMB 1991b).

The GMB maintains that the losses and implied subsidies are inherent in pricing, investment, financing, and administrative policies beyond its control. The requirement for the GMB to operate like a "viable commercial enterprise" while at the same time fulfilling certain noncommercial functions needs clarification, especially relating to the mechanism for reimbursement of its noncommercial functions. Operating competitively alongside private traders requires that the GMB's trading instruments, notably its investments in depot structures and the management of finances, be given a market orientation. Greater managerial flexibility in responding to change in a competitive market environment is essential.

In considering the reorientation of the GMB toward market conditions, the issue to be resolved is "What is the planned role of the parastatal in a post-Economic Structural Adjustment Program (ESAP) Zimbabwe?" Or, "What marketing system do we wish to see in place after the reforms have taken effect?" The type of reforms as well their implementation strategy must be consistent with realizing the stated objectives.

In the GMB, four possible options have been considered for the grain-marketing system in a post-ESAP Zimbabwe: (1) one without the GMB, that is, only the private sector would market grain; (2) one with the GMB providing noncommercial functions only; (3) one with a fully commercialized GMB and vibrant private sector; and (4) one with a vibrant private sector alongside a commercial GMB that has some social responsibility.

The implications for each option must be assessed in terms of how it affects the welfare of the producers and consumers of the products currently handled by the GMB, and also upon the national budget, especially in recognition of the fact that it is the heavy budgetary demands made by the current system that led to calls for its reformation.

Nonexistence of the Grain Marketing Board

Agricultural markets are regulated in virtually every country in the world. One of the major reasons for this regulation is the fluctuations in production and hence in prices. Zimbabwe's small producers are more dependent than commercial farmers on natural conditions. They are also located mostly in the remote, arid parts of the country, away from organized or viable demand for grains. It has been observed in Tanzania that without an effective state organization, liberalization stimulated production in areas better placed to serve major deficit areas such as cities and industrial settlements. Small producers are therefore likely to be kept out of profitable markets by competition from their commercial counterparts and left to satisfy a weather-dependent rural demand for grains. Alternatively, they could still sell in these lucrative markets at very low profit levels, given their high transport costs.

Rainfall has been so erratic that even without changed acreage, maize supplies have fluctuated dramatically. The incidence of dependence on foreign trade for food will increase without the GMB. Even if foreign-trade policy and exchange rates are significantly liberalized, the uniform movement of weather patterns within the SADCC region, similar food tastes of the region, and infrastructural inhibitions to overseas supplies make such reliance on foreign trade for food dangerous.

Profit-driven private enterprises cannot be expected to play the role of market stabilizers or emergency stockpilers without recovering the full costs of doing so. The budgetary subventions required to keep producers on the land or ensure consumers access to affordable food, or both, may not be lower without the GMB than under the current system.

Evidence from some countries that lost their marketing parastatals in the process of reform seems to suggest that wholesale liberalization, managed to reduce food scarcities, and hence prices, in major profitable urban markets but "increased food insecurity in the producing areas and failed to develop an entrepreneurial class capable of undertaking risks" (Amani and Maro 1992). In some cases where small grain producers dominated, it has failed even to generate food for urban areas. Some of the reasons for failing to deliver adequate food to urban areas had nothing to do with pricing, but with capacity to trade. For Tanzania, pooling grain at the village level before transporting it in truckloads to large markets (primary marketing) was particularly difficult. Traders had little capital and handled small volumes, and urban merchants could not guarantee the suppliers an immediate sale on arrival. Small farmers tended to play a passive role in grain marketing and, given the farming community in Zimbabwe, the large-scale commercial farmers would probably take up all the lucrative consumer markets.

The GMB as Provider of "Noncommercial" Functions Only

The GMB could exist solely as a market stabilizer or strategic reserve stockholder. This could be achieved by designating the GMB as a residual buyer or as a holder of buffer stocks, which may double as strategic reserves if made flexible in size.

There is every likelihood that the major profitable markets (commercial supplies and urban industrial demand) would be lost to the parastatal. The commercial sector has the potential to satisfy all viable demand in major consumption areas. In times of good harvests, the remote small producers are likely to continue selling to the parastatal as the thin, weather-dependent demand in those areas withers away. In times of adverse weather, the parastatal would probably sell out in its remote depots. Buyers of last-resort status are likely to restrict the parastatal to buying in remote parts of the country and transporting produce through commercial areas for resale in remote areas and in areas at the extreme opposite end of the country. Distribution costs would rise to levels significantly higher than current levels, and with reduced throughput, unit costs would increase considerably. The financial implications point to higher trading deficits than the current system incurs.

Further, for such market-interventionist policies to be effective, markets would have to operate competitively with few nonprice barriers in the form of transport and information bottlenecks. Given the current oligopolistic nature of industrial processing capacity and the ability of commercial farmers to manipulate supply, limited interventionist policies may not yield the desired results. Instead, random occurrences of poor harvest seasons, which are characteristic of Zimbabwe, may result in GMB stocks being held for much longer than is economically feasible (Theunissen 1987).

Fully Commercialized GMB

A GMB free of nonprofit-oriented types of practices would purchase grain only for profitable domestic and export sales at market-determined prices and would have an infrastructure geared only to performing such profitable business.

A commercial GMB would renounce its development role and the market stabilization and strategic reserve stocking functions. All investments related to these functions would have to be liquidated. High on the liquidation list would be all remote, low-throughput depots. Major urban centers, commercial farms in Natural Regions 1, 2, and 3, and high-productivity communal areas in the same regions are considered by the GMB as its "commercial or profitable markets." The parastatal would operate as another private trader and the implications for the various clientele groups (commercial and small producers, consumers, and the state budget) would be the same as those implied by the nonexistence of the GMB.

If the goal is to have a commercial GMB in the post-ESAP Zimbabwe, reform efforts must center on the following: (1) systematic pruning down and reorientation of GMB infrastructure, especially its depot network, toward market dictates; (2) alignment of pricing structure with market forces; (3) financial reforms to allow the parastatal complete autonomy in managing its finance portfolios; (4) phased relaxation of restrictions on grain movement; and (5) sharpening the analytical tools within the parastatal so that it meets the challenges of competitive trading; quantitative analytical capacity, especially relating to investment appraisal, pricing analysis, market intelligence, and strategic planning have been identified in recent studies as main areas of weakness in the board (Coopers and Lybrand 1987; Food Study Group 1990).

A Commercial GMB with a Valid Social Role

In the reform program announced by the government, it is stipulated that the GMB will continue to provide the noncommercial service of market stabilization. The reform document (Zimbabwe 1991) is silent on strategic reserves and development, but in successive discussions with government officials, it appears that was an inadvertent omission. It was these discussions that led GMB management to conclude that the government intended to designate the GMB as a commercial organization with a valid social role.

As argued above, setting up an organization solely to perform market stabilization and to hold strategic resources is likely to be a costly move, considering the geographic dispersion of production, weather factors, and general infrastructural bottlenecks to trade. A commercial GMB would be able to subsidize the noncommercial functions with profits from the commercial account so that the demand on the treasury would be reduced.

The reform efforts must, in the short term, concentrate on improving the efficiency with which the GMB performs its commercial functions and the effectiveness with which it performs the noncommercial functions. For the medium-to-long term, the reforms must aim at developing sustainable marketing systems that allow both private and public agents to play a role in grain marketing. Of particular importance is the maintenance and strengthening of access by small producers to viable low-volume, incentive-generating markets for their produce and low-cost distribution of inputs to this sector. The development of appropriate exchange-rate policies, so that foreign exchange becomes available for importers of inputs, vehicles, processing equipment, and spare parts "is an important component of a favorable environment for entrepreneurial skills to flourish" (FAO 1990).

The separating factor between the above options appears to be the sequencing of the reforms and the pace at which they are implemented. As argued above, direct competition between the GMB and the private sector that is not preceded by major policy reforms on pricing and management flexibility will result in a clear separation of markets between high-cost markets that will be serviced by the GMB and low-cost markets that will be serviced by the private sector.

The GMB related the high cost of performing noncommercial functions to the absence of small-scale, private-sector traders in communal lands and the chronic weakness of the rural transport system. The GMB is forced to provide services to a sector whose mode of operation and scale of business is not compatible with the economies of scale reminiscent of a large organization. The success or failure of the reform program will depend on whether a strong rural trade in grains emerges, either to trade in the small markets as an end in itself or to bulk up for bigger markets in urban centers and the reserve. The GMB recommendations are combined in Table 10.1.

Table 10.1—Summary of phased partial maize reform program

Reform	Stage 1, 1992/93	Stage 2, 1993-95
Regulatory changes	Free intercommunal area movement Permission for approved buyers to trade locally Seminars to educate local officials on new, partly liberalized system	Free movement to all areas except zone centers Small-scale movement allowed into zone centers (five bags per vehicle) Zone center millers to buy from GMB Deregulation of other millers' sourcing
Pricing policy	Introduction of differential for off-rail producer price (partly subsidized) Producer price innovations to ensure stable supply to GMB	Introduction of flexible ex depot GMB selling price in all areas outside zone centers Removal of price controls on market transactions except in zone centers Review of producer transport subsidy
Institutional development	GMB to contract with major suppliers Seasonal closure of some depots Five depots to close Formal performance contract between government and GMB	GMB to contract with local millers GMB to introduce commercial storage service GMB to lease depot space Continued rationalization of depot network
Investment	Identification of priority areas for feeder road development	Investment on basis of study recommendations Investment in physical grain-trade facilities in small towns
Market promotion	Policy statement on role of grain trader Establishment of market-information collection and dissemination system Study of constraints to private trade Establishment of bank credit programs	Continuation of previous actions
Promotion of processing	Review of constraints to peri-urban hammer milling	Urban councils to designate areas for hammer mills Easing of regulatory constraints to peri-urban hammer milling
Monitoring and evaluation	Design of national monitoring and evaluation system	Implementation of national monitoring and evaluation system

Source: Grain Marketing Board, Response from the Management of the Grain Marketing Board to the Economic Structural Adjustment Program (GMB, Harare, 1991).

THREATS TO GRAIN MARKETING REFORMS

The reform program is threatened by the lack of a clear policy direction and the lack of consultation and coordination. The uncertainty of how grain liberalization will impact on small producers is a major concern. A strong antitrader ethos among small communal producers who regard traders' constraint-driven behavior as exploitative promotes pressure for maintenance of the status quo.

There appears to be some belief among policymakers that major cost savings could be obtained by merely improving the operational or managerial efficiency of the GMB. This takes pressure off the real policy issues behind the problems in agriculture.

Large-scale producers feel they have been discriminated against for too long, and preoccupation with pricing has resulted in sidelining issues of productivity on the farm, bottlenecks in the distribution of inputs, the exchange-rate effects, and the general macro policy impact on agriculture.

Donor pressure could lead to unsynthesized action plans at unsustainable places. In a country that has lived under controls since 1933, it is important that changes in the regulatory environment be announced and publicized both before, during, and after their implementation.

CONCLUSIONS

The need for liberalizing grain marketing is generally recognized, and therefore the reform program is welcome, as it should offer a number of benefits including the elimination of impediments to increased growth in grain production.

In carrying out the reforms, a proper balance should be struck between commercial prudence and guarding the national interest, which is not necessarily expressible in monetary terms, including food security and development. The hard-earned gains made by the small-producer sector must be adequately protected to ensure that this sector continues to contribute significantly to the economy. This is essential not only in terms of rural development but in terms of keeping rural people on the land and reducing rural-urban influxes that could easily negate the whole structural-adjustment program. To achieve that, the reforms must allow for a phased withdrawal of protection of the GMB.

Pricing-policy reform is the dominant issue in the reform program and the price reforms should go hand in hand with increased flexibility for both private and parastatal management.

Infrastructural constraints to private trade have been underplayed, and evidence from countries that have liberalized without paying sufficient attention to this issue is not encouraging. Both producers and traders have responded to the reforms much slower than anticipated "due to infrastructural, sectoral, and macroeconomic constraints" (Santorum and Tibaijuka 1991).

Well managed, the reforms should not be excessively expensive given that the main actions relate to regulatory and pricing-policy changes. However, demands on the treasury are likely to be increased by the call for the improved rural communication network that is crucial to development of a vibrant rural grain trade.

REFERENCES

Amani, H. K., and W. R. Maro. 1992. Stock management problems and policy under market liberalisation for grains in Tanzania. In *Food security research in southern Africa: Policy implications*, ed. J. B. Wyckoff and M. Rukuni, 192-206. Proceedings of the Seventh Annual Conference on Food Security Research in Southern Africa, 28-30 October 1991, University of Zimbabwe/Michigan State University Food Security Research Project. Harare: University of Zimbabwe, Department of Agriculture, Economics, and Extension.

Coopers and Lybrand Associates. 1987. Improving operational efficiency of the grain Marketing Board. Confidential report. Grain Marketing Board, Harare. Mimeo.

FAO (Food and Agriculture Organization of the United Nations). 1990. Selected papers on structural adjustment and grain marketing, Marketing and Credit Service Division. Rome. Mimeo.

Food Study Group. 1990. *Agricultural marketing and pricing in Zimbabwe*. Background report for the World Bank. Zimbabwe Agricultural Sector Memorandum. Oxford, U.K.: Queen Elizabeth House, Oxford University.

GMB (Grain Marketing Board). 1991a. Proposals for deficit reduction. Background analysis to GMB Business Plan. GMB Planning Unit, Harare.

_____. 1991b. Response from the management of the Grain Marketing Board to the Economic Structural Adjustment Program. GMB, Harare.

_____. Various years. *Annual report*. Harare: GMB.

_____. Various years. *Annual trading accounts*. Harare: GMB.

Jayne, T., M. Rukuni, M. Hajek, G. Sithole, and G. Mudimu. 1991. Structural adjustment and food security in Zimbabwe: Strategies to maintain access to maize by low income groups during maize marketing restructuring. In *Toward an integrated national food policy statement*, ed. J. Wyckoff and M. Rukuni, 8-50. Proceedings of the Second National Consultative Workshop, 10-12 June. Harare: University of Zimbabwe, Department of Agricultural Economics and Extension.

Jiriyengwa, S. 1989. An econometric analysis of the GMB maize trading account. Paper presented to the M.Sc. class, Department of Economics, University of Zimbabwe, Harare.

Muchero, M. 1986. A preliminary analysis of the causes of GMB losses. Internal papers of the Grain Marketing Board, Harare.

Santorum A., and A. Tibaijuka. 1991. Trading responses to food market liberalization in Tanzania. Paper presented at the Workshop on Trading Responses to Food Market Liberalization in Sub-Saharan Africa, 2-4 October, University of Zimbabwe, Harare.

Theunissen, T. 1987. *Grain marketing and food security in Zimbabwe.* Oxford, U.K.: Food Study Group, Queen Elizabeth House, Oxford University.

USDA (U.S. Department of Agriculture). 1991. *Targeted consumer food subsidies and the role of U.S. food aid programming in Africa.* Conclusions of a Workshop held on 21 November 1989. Workshop report document no. PN-AGB-831. Washington, D.C.

World Bank. 1991. Zimbabwe agricultural sector memorandum, vol. 2. Washington, D.C.: World Bank, Southern Africa Department.

Zimbabwe, Ministry of Finance, Economic Development and Planning. 1991. *A framework for economic reform (1991-95).* Harare: Government Printer.

11 Implications of Regional Trade Liberalization for SADCC's Food-Security Program

Roger W. Buckland

Most Southern African Development Coordination Conference (SADCC) countries are primarily agrarian, but a declining per capita food production has deepened dependence on food aid in the region over the past two decades. Most of the countries have not been able either to produce sufficient food for themselves or to generate enough foreign exchange to ensure that the food needs can be imported.

Recognizing that their own policies have contributed to this, many SADCC countries have introduced macroeconomic reforms to reduce fiscal and monetary imbalances and to promote sustained growth in investment, output, exports, and employment. Deregulation of domestic markets will be introduced to allow product and labor prices, among others, to move in line with demand and supply and to remove parastatal inefficiencies. Trade liberalization is usually an important part of the comprehensive structural adjustment programs.

When the Lusaka Declaration that established SADCC was signed in 1980, a major objective was to accelerate economic growth, including regional cooperation to improve regional food security. A broad program to achieve regional food security was formulated by the mid-1980s. Significant elements of the original program remain relevant today, particularly the stimulation of agricultural output. However, increasing emphasis will have to be given to programs that are aimed at households rather than at countries as a whole. In particular, promoting access to food will have to receive much greater attention, as will increasing information and commodity flows across borders.

SADCC's food-security program is already being adjusted to take account of trade liberalization. More emphasis must now be placed on ensuring food security for vulnerable groups. Conservation of the region's soil and water resources at a time when rising populations are putting increasing pressure on them must also receive more attention.

ECONOMIC BACKGROUND

SADCC now comprises 10 member states, covering an area of 5.7 million square kilometers. Its population is estimated at nearly 86 million, its gross domestic product is about US$28 billion, and average per capita income is US$330 (Table 11.1).

Despite the geographic proximity of SADCC countries, their economies differ substantially. Reliance on agricultural exports is far lower in Angola, Namibia, and Zambia than in most other SADCC member states, yet over two-thirds of the work force in those countries relies on the agricultural sector for a livelihood. Clearly, the contribution of agriculture to the economy is far less for Botswana than for many other member states. However, in common with the rest of the region, the bulk of Botswana's population is still employed within the sector. What happens in the agricultural sector of each country, therefore, is critical to the livelihood and food security of a large proportion of the country's population.

Table 11.1—The economic importance of agriculture in SADCC countries, 1989

Country	Population	GDP	Share of GDP from Agriculture	Share of Labor Force in Agriculture	Share of Exports from Agriculture
	(millions)	(US$ million)		(percent)	
Angola	10.0	7,720	20.0	71.0	3.7
Botswana	1.3	2,500	3.0	65.2	8.4
Lesotho	2.0	340	21.0	81.7	67.6
Malawi	8.4	1,410	37.0	77.7	93.3
Mozambique	15.0	1,100	62.0	82.5	52.6
Namibia	2.0	1,650	10.7	77.2	11.0
Swaziland	1.0	369	39.5	68.7	57.6
Tanzania	28.3	3,080	66.0	82.3	79.3
Zambia	8.0	4,700	14.0	70.1	3.5
Zimbabwe	10.0	5,250	11.0	69.6	45.6

Source: Data for population and total GDP are from World Bank, *World Development Report* (Washington, D.C.: World Bank, 1991); SADCC, *Regional Food Marketing Infrastructure Study* (Harare: SADCC, Food Security Unit, 1989).

Note: SADCC is Southern African Development Coordination Conference.

Total merchandise exports from SADCC countries tend to be limited to one or, at best, very few commodities. One mineral (this varies from country to country) dominates the exports of Angola, Botswana, Namibia, and Zambia. For the other SADCC countries, five commodities (tobacco, coffee, cotton, sugar, and tea) account for over 75 percent of SADCC's total agricultural export earnings (Table 11.2).

Terms of trade for primary product exports relative to manufactured imports have been declining since the late 1970s, and many continue to fall in the 1990s. With declining real prices and little or no growth in volumes, the region's real export earnings actually fell by 33 percent during the 1980s (Table 11.3).

SADCC'S FOOD-SECURITY PROGRAM

At SADCC's inception, priority was given to programs that promoted regional cooperation in economic development. Food, agri-

Table 11.2—SADCC dependence on major agricultural commodities for agricultural export earnings, 1987

Country	Tobacco	Tea	Sugar	Cotton	Coffee	Subtotal
			(percent)			
Angola	0.0	0.0	0.0	4.6	88.7	93.3
Botswana	0.2	0.0	0.0	0.0	0.3	0.5
Lesotho	0.0	0.0	0.0	0.0	0.0	0.0
Malawi	68.1	11.1	8.9	0.1	3.7	91.9
Mozambique	0.5	0.9	5.7	12.7	0.0	19.8
Swaziland[a]	0.0	0.0	66.1	3.5	0.0	69.6
Tanzania	5.4	5.1	1.5	14.5	45.9	72.4
Zambia	45.2	0.0	25.7	2.3	8.3	81.5
Zimbabwe	50.0	1.9	9.7	13.9	4.4	79.9

Source: Based on data from World Bank, "Agricultural Diversification and Trade in SADCC Countries" (World Bank, Washington, D.C., 1991, mimeographed).

Notes: SADCC is Southern African Development Coordination Conference. Mozambique's leading exports are prawns and cashew nuts. The agricultural exports of Botswana and Lesotho are dominated by meat and mohair, respectively. Data are not available for Namibia.

[a] Average for 1984-86.

Table 11.3—SADCC's agricultural export earnings, selected years

Country	1980	1985	1988
		(US$ million in 1980 prices)	
Angola	165.9	75.5	20.9
Botswana	44.3	81.2	20.9
Lesotho	14.7	16.7	7.9
Malawi	240.6	232.3	197.2
Mozambique	150.4	37.8	40.3
Namibia	94.3	58.2	116.8
Swaziland	202.3	102.5	118.0
Tanzania	391.3	250.0	185.0
Zambia	12.9	17.4	11.5
Zimbabwe	434.5	494.2	398.0
Total	1,751.2	1,365.8	1,116.5

Sources: SADCC, *Regional Food Marketing Infrastructure Study* (Harare: SADCC, Food Security Unit, 1989); World Bank, "Agricultural Diversification and Trade in SADCC Countries" (World Bank, Washington, D.C., 1991, mimeographed).

Note: SADCC is Southern African Development Coordination Conference.

culture, and natural resources were given special attention. The broad objectives of the program were to provide a framework for integrating SADCC's national and regional policies relating to agriculture and natural resources.

Two broad strategies to achieve the objectives were prepared in the mid-1980s. These are currently being reviewed and reformulated into a single coherent program. The component sectors of the program are food security, agricultural research, livestock production and disease control, forestry, inland fisheries, marine fisheries, wildlife, and environment and land management.

The principal aims for the food-security sector are to integrate national and regional food-security policies, promote increases in food and agricultural output, help eliminate periodic food crises, and develop programs aimed at raising rural incomes, generating rural employment, and improving household food security.

The strategy to achieve these objectives includes (1) developing a mechanism for exchanging technical and economic information related to food security; (2) reinforcing national food-production capacities; (3) improving food storage, distribution, delivery, conservation, and processing systems; (4) promoting diversification into cash crops and agro-industrial enterprises; (5) establishing systems for preventing food crises and developing national food-security strategies; (6) establishing programs to control major crop pests and crop diseases; (7) developing skilled manpower; and (8) developing intraregional trade.

SADCC's program for achieving regional cooperation in food security has involved a step-by-step approach, rather than one that requires a high degree of integration of individual states' policies from the outset. The current regional food-security program is designed to complement member states' national policies. It has never been the role of the food-security program to be involved directly at the national level, except where regional projects have national components. Production policies at the national level are the prerogative of each state, so SADCC's contribution has usually been in the form of providing information, assisting in the research effort, improving marketing within the region, and trying to reduce the impact of seasonally induced shortfalls of staple foods.

These indirect inputs, especially in the form of research, training, and project planning, have been substantial. Liberalization now provides the opportunity for SADCC's regional food-security program to evolve into one that has a much more direct facilitative role in improving the availability and access of households to food.

EFFECTS OF ADJUSTMENT MEASURES

A major aim of adjustment programs is to make national economies more competitive. Common elements of many such programs are relaxation of market controls, reductions in fiscal deficits, and devaluations. These measures, together with liberalization of foreign trade, are designed to allow economies to benefit from comparative advantage and to become more integrated with world markets, thereby fostering agricultural development and, implicitly, encouraging an increase in overall output as well as diversification. If diversification into other crops is widespread, some states will have to exhibit a greater willingness and capacity to rely on external sources of food. Alternatively, productivity increases in staple foods will have to be even faster than is required by increasing population.

An implicit assumption underpinning the adoption of liberalization policies is that the private sector is more efficient in allocating scarce resources and hence in promoting development. However, less national control also implies that regional programs that facilitate the growth of agricultural development and trade will have an enhanced role. Under more liberal regimes, the agricultural sectors of individual member states could become much more closely interlinked. Governments will not be able to rely on regulation or to access subsidies to the same extent to achieve sectoral objectives. So development policies in each state will have to take increasing cognizance of the trading conditions and policies of other states. SADCC's regional programs will need to adjust to these developments.

A further assumption is that reform will result in improvements in output and hence in food security. An improvement in producer incentives and in the efficiency of markets may well elicit a supply response. But liberalization has implications for both sides of the food-security equation, not just the supply side. Theory suggests that freeing resources for higher marginal value uses should yield a net benefit to an economy. However, this may well make some groups worse off than before. Distributional issues are obviously important in this context. Policymakers will also need to take account of the uses to which the freed resources are put and whether those uses are environmentally acceptable.

Social (health, education) and regional development programs are frequently major elements of government expenditures. When government spending on these programs is reduced, it raises the issue of knowing who is affected, where they are, and what is likely to be the best strategy to ensure greater food security. Clearly, improvements in the region's data base would help ensure that food-security programs are targeted effectively.

Curbing fiscal deficits can also result in less government spending in rural areas. Reduced access to subsidized markets will make the sale of high-bulk, low-value commodities (for example, maize) unprofitable for producers in more remote surplus areas. In addition, withdrawal of some government services from rural areas will reduce real incomes. SADCC's food-security program, therefore, will need to include strategies to promote rural employment, including, for example, working through nongovernmental organizations.

The effect of structural adjustment on food security can be summarized as follows:

- It can increase (reduce) incentives to farmers who produce marketable surpluses (in this case, the effects are felt on the supply side of the food-security equation).

- It can reduce consumer prices (demand side of the equation) in urban and rural areas if marketing margins are reduced, thereby effectively raising the purchasing power of current incomes; where subsidies are in place, however, prices could rise.
- It is felt differently by wage earners, especially the urban poor, compared with agricultural producers in general.
- The effects can also vary substantially within a given group; farmers who produce a surplus can benefit if prices rise, while those in deficit are disadvantaged, and differences in resource levels (physical as well as human) can show up as large variations in the capacity of individual households to adjust to changes in input and output prices.
- The characteristics of groups at risk of food insecurity will also change; when government intervention is reduced, some producers in remote areas may find farming less profitable (for example, those a long way from markets but who previously received the same price as those nearer to markets), and fixed-income consumers may also have to face higher prices.

CHANGING FOCUS

SADCC's food-security program has focused primarily on helping member states to increase food availability through expanded domestic production, reduction in losses, and improvements in national and regional storage. However, there is ample evidence to suggest that increases in food production, even to the point of meeting total national dietary requirements, do not automatically end hunger and malnutrition. The realization that hunger continues to coexist with plenty means that attention is now focused much more closely on producing and consuming units, that is, households (Jayne, Wyckoff, and Rukuni 1990).

This is not to say that regional and national food self-reliance (that is, encouraging increases in output) will no longer figure in the sectoral program. It is just that greater emphasis will now be given to household food-security issues and the implications they hold for regional strategies. This focus gains added importance under structural adjustment. Not only must we try to reduce the social costs of adjustment, the regional program should aim at providing the framework for facilitating improvements in productivity at the household level that can be sustained over the long term.

More emphasis will also be put on the demand aspects of food security. A household's capacity to acquire food is influenced by a whole complex of factors. Programs to stimulate regional agricultural development may well be the most effective way of fostering rural employment and higher rural incomes. Even then, distributional problems will inevitably remain. Nonetheless, relaxation of trade and other regulations now provides an opportunity for further detailed research and more open discussion of such policy issues.

Apart from the need to reframe the food-security program, increasing pressure on land is forcing the issues of productivity increases, prevention of environmental degradation, and rural employment onto the agenda for action. Continuing population growth of around 3 percent per year implies that SADCC's agricultural output will have to rise by at least that amount merely to retain present levels of self-reliance, and by a further 1 percent if modest improvements in average incomes are to be attained. Past increases in output have tended to come largely from the expansion of cropping into progressively more marginal land and more intensive use of existing arable land. The effects of this have been to reduce the area available for grazing, to subject the more-fragile environments to cropping (with the inevitable losses of soil and of crops), and to exhaust the soils in some areas because the rotations normally practiced in traditional croplands are no longer possible.

Natural-resource conservation is therefore an essential element of the food-security program. Increased rural processing, crop diversification, and development of appropriate land-management regimes all have a role to play. Strategies to promote productivity increases and soil and natural resource conservation are therefore being integrated into the regional food-security program.

New Program Aims

Given the reduction in intervention at the national level, SADCC's food-security program should now be aimed at (1) providing additional information, food policy analyses, and research (particularly on vulnerable groups), which will involve building up the capacity to collect data at the national level as well as in the regional unit; (2) facilitating the formation of a regional capacity to respond to short-term shortages of major food staples (a revised food reserve project); (3) promoting farming systems that conserve soil and water resources while maintaining food security even in the short term (this will necessitate continued research into cropping systems as well as more efforts to incorporate agroforestry and livestock systems); (4) promoting rural processing and other small-scale employment, with special regard to women; (5) providing a forum for regional cooperation in food trade through improvements in marketing infrastructure and information flows; (6) promoting regional development of water resources and an increase in irrigation of staple food crops as well as cash and export crops; and (7) facilitating the mobilization of assistance to member states to enhance training.

STRATEGIES

As indicated earlier, many of the strategies for improving regional food security remain as relevant in a liberalized environment as they were before. Improving productivity, which was always seen as important, will remain so. Market, rather than administered, prices may now change the rate of productivity growth and could change cropping patterns. The question is how to promote higher productivity, retain natural resource bases, and improve household food security in general.

Research and Extension

Expanding populations, increasing pressures on land, the need to conserve fragile environments, and the importance of introducing cropping systems that enhance food security without environmental damage all point to a need to encourage further agricultural research. Climate variability and the push into more marginal cropping areas are important determinants of the direction such research should go in. But equally important, the regional food-security program should provide some means to strengthen extension efforts at the national level and to have the benefits of this research introduced at the farm level.

Policies aimed at strengthening individual economies will inevitably have to be reviewed and adjusted. The impact of adjustment packages on particular segments of the population, and on food security in particular, implies a need for increased economic-policy research. Skills that exist within the region could be brought together under some form of network to contribute to the development of appropriate regional or sectoral policies. Being able to put policies in place that really are effective in improving food security implies that considerable research precedes their adoption. An important ancillary requirement, of course, is an improvement in the availability and relevance of data.

Fashioning schemes to provide a safety net for vulnerable groups, or indeed to help them develop economically, means that the existence and the characteristics of such groups have to be known. So there is a need for better socioeconomic data. Promoting trade and rural income generation also implies a need for good statistics.

The SADCC Food Security Technical and Administrative Unit has developed a proposal for a regional policy-analysis network that would draw on regional expertise to provide a forum for economic debate and formulation. It is envisaged that it might also provide an input into policy-analysis training.

Information

More regional data are likely to be required in the aftermath of liberalization. Output of major staples and expected requirements to meet nutritional standards and regional availability will need to be known for forward planning and for food aid. This implies a need to maintain and strengthen the Regional Early Warning System with all its national components.

The characteristics and prevalence of groups vulnerable to food insecurity also need to be known to formulate relief programs. A nutrition-monitoring project is already being developed within the food-security program. The demand characteristics of households are also necessary in order to devise appropriate regional trade and development policies. Further detailed economic research into questions such as this will necessitate substantial improvements in data. Promotion of information on household storage and processing techniques is also needed to enable individual households to reduce seasonal food shortages, to reduce waste and raise the nutritional inputs from a wider range of crops, and to promote avenues for earning cash incomes.

In order to promote market integration, knowledge of domestic, regional, and world market prices is important. The regional information-system project is seen as the appropriate vehicle for achieving this.

Regional trade in staple grains could be facilitated through the provision of trade data, including continuously updated information on local prices, freight rates, and availability. The Food Security Technical and Administrative Unit has an information section within its overall structure that could undertake this role.

Regional Training

The main element here would be to enhance the region's agricultural policymaking and project-management capacity, probably in conjunction with the African Capacity Building Foundation. The concepts of networks, of using regional institutions (both the Human Resources Development sector of SADCC and regional tertiary institutions), of ensuring sustainability, and of constant appraisal will all need to be incorporated. The major areas of concentration are currently identified as crop marketing, transport, and storage, particularly of grain, but also of cash crops and export commodities; project preparation and management; crop protection; and policy analysis.

CONCLUSION

SADCC is facing a number of challenges. High rates of population growth and the need to stimulate stagnating economies to maintain, let alone increase, per capita incomes have resulted in the implementation of adjustment programs in many SADCC countries. These, together with changes in world economic conditions, have prompted a reappraisal of the region's food-security program. A new program is evolving that focuses on individual households to a much greater degree. Further, demand aspects of food security will now figure much more prominently in the development of the program.

Trade liberalization means that there are new opportunities for a regional body to contribute. As freer trade in agricultural commodities becomes possible, data requirements are bound to increase. An expansion of SADCC's capacity in this regard is already in hand. Closer coordination of policies will also be necessary. The program could contribute by facilitating the discussion of the policy implications of food-security issues in the region. If possible, the present capacity to undertake agricultural and policy research must be increased. More human and financial resources are needed for both these areas of endeavor. Training and capacity-building are therefore important components of the new program. Marshaling resources to enhance the region's marketing infrastructure will also be necessary.

The capacity of the region to raise productivity without putting its future in danger by degrading its resources poses one of the most important problems for the future. Fostering the use of farming systems appropriate to specific areas and developing policies that encourage their use, thereby ensuring a capacity to produce staple foods in sufficient quantities over the long term, will therefore be a major issue in the evolving food-security program.

REFERENCES

Jayne, T. S., J. B. Wyckoff, and M. Rukuni, eds. 1990. Integrating food nutrition and agricultural policy in Zimbabwe. SADCC/UZ/MSU Food Security Project, University of Zimbabwe, Harare.

SADCC. 1989. *Regional food marketing infrastructure study*. Harare: SADCC, Food Security Unit.

World Bank. 1991a. Agricultural diversification and trade in SADCC countries. World Bank, Washington, D.C. Mimeo.

World Bank. 1991b. *World development report*. Washington, D.C.: World Bank.

12
Price Adjustments and Distributional Consequences of Trade in Maize in Malawi, Zambia, and Zimbabwe

Ernst-August Nuppenau

Trade in agricultural products between neighboring countries is rare in Sub-Saharan Africa. Due to government regulations on international and intraregional movements of foodstuffs, trade in agricultural commodities is limited to special products only. In principle, governments follow different strategies of autarky in food markets in order to prevent their populations from becoming dependent on foreign markets. In the southern African context, where marketing boards regulate food marketing to a considerable degree, domestic food markets are insulated from markets in neighboring countries. Currently, Malawi, Zambia, and Zimbabwe follow a strategy of food self-sufficiency in normal years. Since it is expected that these countries will not have identical patterns of comparative advantage, a policy of self-sufficiency will inflict economic costs. There is evidence that these countries could engage in considerable trade in agricultural commodities with each other (Koester 1990). It is the objective of this study to show how trade between these three countries would change maize quantities, prices, and distribution positions.

Of course, since drought is prevalent in the region, protecting national populations from food shortages is a major concern. One means of increasing food security is carryover stocks, which in Zimbabwe, for example, have reached the level of one year's consumption, or 1.2 million metric tons, in some years. Since such stocks are a very costly means of insurance, other measures to ensure food security, such as trade and reduced stockpiling, may become options in the future. Governments in the region are becoming increasingly aware that autarkic policies place heavy burdens on government finance, immobilize capital permanently, and slow down development in other sectors. Koester (1986) suggests that intraregional trade should be investigated as an option for improving food security.

This contribution can be seen as an attempt to quantify the effects of the introduction of maize trade in the SADCC region. Obviously, liberalizing trade requires changes in the institutional setting. For example, private traders could assume responsibilities that are currently borne by parastatals and marketing boards. Liberalization would lead to price differences, which are the engine of arbitrage and thus of trade. Of course, the same results could theoretically be generated under parastatal trade if margins for transport, stockpiling, and other activities indirectly involved in marketing grains were included.

Simulations will provide insights into various scenarios and conditions; in particular, the effects of drought on trade in the region and a change in import routes. A change in import routes may occur due to political constraints on trade involving South Africa. These simulations provide the basis for a more detailed analysis and discussion of the results of trade liberalization. Trade liberalization is already on the agenda in all three countries and policymakers need information on the particular effects of their policies.

The study is structured in three sections. The first section provides a formal introduction to the basics of an interregional spatial-equilibrium model. In the second section, the empirical background that serves as a baseline for the discussion of changes in the maize economy due to changes in trade is outlined. In the third section, results from various model runs are presented.

MODEL-BUILDING

The analysis requires a framework that takes into account various interactions between subregional production and consumption. Interaction or its potential emerges in the form of trade activities created by price differentials between subregions. Under free trade, subregional prices will adjust according to transport costs between subregions. Such conditions can be modeled using spatial/partial equilibrium models.

A description of the formal framework follows. The pioneering work can be attributed to Takayama and Judge (1971) and the approach in the present study is closely linked to Hazell and Norton's (1986) description of sector models for investigating spatial market equilibria. In principle, the model operates with the objective function of maximizing consumer surplus minus production, transport, and stockpiling costs. Stockpiling costs are relevant, because a two-year model is used to simulate stockpiling in the first year in anticipation of a drought in the second year. Producer surplus is automatically included by the calculation of subregional prices and production levels and costs. The quadratic objective function is designed according to standard approaches (Bale and Lutz 1981). Demand and supply functions are assumed linear in order to keep the approach as simple as possible. This may become a problem if large changes occur, but serves as a first-order approximation. In detail the model can be described by a simplified diagram (Figure 12.1) that comprises three subregions and displays the initial approach in a normal year.

Normal Year

For the sake of simplicity the model description starts with only two countries: country 1, which includes both a surplus and a deficit subregion, and country 2, which includes only a surplus subregion. In Figure 12.1, a situation with no price differentiation between subregions in country 1 (panterritorial pricing) and no trade with country 2 is compared with an alternative situation where prices reflect transport costs and trade occurs between and within countries. To compare these two situations, distribution and welfare effects have to be considered separately. In the situation with panterritorial pricing, the government incurs transport (budget) costs in country 1. Since there is no trade, both countries are forced to be 100 percent self-sufficient. Thus price p10 prevails in country 1, and price p20 prevails in country 2. Abolishing panterritorial pricing in country 1 and allowing trade between the two countries, the situation changes to an integrated market depicted by the dotted price and quantity lines. In the case of trade, transport costs matter. Individual prices p11 and p21 in the surplus

Figure 12.1—Model framework to evaluate trade liberalization in a spatial-equilibrium model

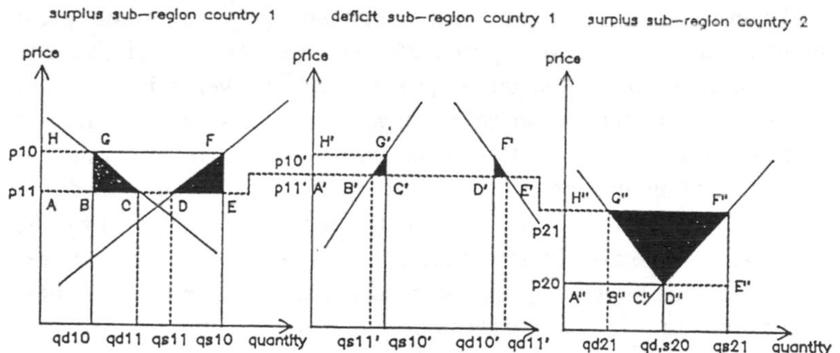

subregion depend on its distance to the deficit subregion. If country 2's surplus subregion is relatively far away, its price level will be lower than that in country 1's surplus subregion. Of course, this need not be the case. As shown later, the Zambian capital, Lusaka, for example, is closer to the Zimbabwean Mashonaland West than it is to the surplus subregion in Zambia's Southern Province.

From a theoretical point of view, the model result does not change substantially if international trade is introduced into the model. International trade can be seen as trade with a further subregion that happens to be characterized by an infinitely elastic supply function at the world market price.

In Figure 12.1 the price p11 is lower than p10 and p11 is less than p10′. The opposite takes place in country 2, which now exports to country 1. However, price increase in country 2 is softened by the transport-cost difference. From a distributive point of view, the main losers on the producers' side will be found in country 1, where producers of the now-tradable commodity no longer enjoy subsidized procurement. Similarly, consumers in country 2 will lose their privilege of low food prices in favor of consumers in country 1. However, country 2 as a whole will gain from trade because producers make an extra profit from financial surplus obtained by exports. This financial surplus is represented by the area B″E″F″G″.

As depicted, consumers in country 1 will enjoy lower prices than in the initial situation. However, it is not possible to predict in a qualitative way which will be the cheapest surplus subregion for a given deficit subregion. Only the interaction of all subregions within the model can provide such predictions. Specifically, the question to be answered is, Which surplus subregion at what minimum price will anchor the overall price structure? Even in the situation of trade with third countries, the question of who will be the region's minimum-price supplier remains. Since transport costs vary between subregions, only the solution of an optimization process can answer this question. Furthermore, the model can only indicate the effects for a specific set of data. Solutions may differ from year to year. Hence, only an illustration of an average over the years 1985-87 can be provided.

The overall gains from trade for the region are indicated by the shaded areas. The distribution of these gains has been described above. An uneven distribution of gains can be detected by the model presented in this study and can have an important impact on the prospects for successful liberalization. For example, if country 2 obtains major gains from trade relative to country 1, some mechanisms for sharing these gains may be required in order to persuade country 1 to liberalize.

Drought Year

As mentioned above, drought is a major problem in the SADCC region. Since drought has unequal effects on subregional supply, price structures must adjust to restore market equilibrium. In the model applied here, adjustment takes place in various consumption areas simultaneously. A crucial question is whether trade or stockpiling or both can alleviate production shortfalls by spreading the effects over more participants. In order to investigate these questions, a two-period model of trade and stockpiling has been designed. This model adds to the one-period model by simulating drought conditions in the second period. Stocks built up in the first period are allowed to alleviate shortfalls in the second period. It is assumed that these stocks are built up in the anticipation of typical drought conditions. Price differences between the two periods guarantee the profitability of stockholding. Note that the question of private versus government stockpiling is not addressed here.

This approach is deterministic in the sense that the shortfall is anticipated and only certain typical drought episodes based on past probabilities are investigated. In reality, the question of strategic stockpiling involves risk aversion, expectations, and other interactions that are beyond this limited approach.

The simulation's results (prices, quantities consumed and produced, and trade volumes) are presented in a later section. In order to understand these results, however, it may help to take a brief look at baseline conditions and the parameters that have been used.

EMPIRICAL BACKGROUND

Surpluses and Deficits by Subregions

It is not very easy to get reliable information on production and consumption in African countries for the purpose of long-run planning. However, some national studies contain recently updated statistical information. In order to make the data comparable, the year 1986 was chosen as the baseline. Average production from 1985 to 1987 was calculated in order to avoid one-year production fluctuations. However, the exact identification of surplus and deficit areas is not crucial for this study, which serves as an illustration. On the other hand, empirical observations guarantee that the illustration is linked to a current situation.

Note that areas with the highest production often coincide with areas of large surpluses (Figure 12.2). Furthermore, apart from some peripheral regions, especially in Zambia, deficits are concentrated in agglomeration centers. However, some rural areas can very quickly become deficit areas, as can be demonstrated by simulating drought situations.

Subregional supply and demand figures are used as benchmarks for constructing linear supply and demand functions. To do this, information on elasticities of supply and demand is required. Since no reliable subregional elasticities have been reported in the literature, standard values are assumed. Throughout, the paper assumes a demand elasticity of -0.4 and a supply elasticity of 0.85 for commercialized subregions. For communal or peasant-dominated subregions, a supply elasticity of 0.6 is applied. These figures coincide with figures from a study by Buccola and Sukume (1987), who report a national demand elasticity of -0.4 and a supply elasticity of 1.0 for Zimbabwe. However, in order to make it a conservative estimate, a 40 percent reduction in responsiveness has been anticipated. Again, this approach coincides with a first-order partial liberalization approach as applied in the whole study. Baseline nominal panterritorial prices of US$71 in Malawi, US$110 in Zimbabwe, and US$133 in Zambia are used (Valdés 1990).

This information is a prerequisite for the design of subregional supply and demand functions in a normal year. A normal year includes adjustments in supply and demand. In contrast, drought years include changes in demand only. Furthermore, in order to keep the approach simple, no parallel market activities are introduced. This implies that production and consumption decisions have been made on the basis of official pricing—an assumption that might not reflect reality.

Routes

Transport costs are drawn from an estimation by Berger (1986) that may underestimate prevailing costs in the late 1980s, given the deepening crisis in the regional transport system. New model runs might be appropriate if improved data becomes available. Transport costs applied in the model are listed in Table 12.1.

Assessment of the Reliability of Production in the Case of Drought

Looking at the issue of food security and intraregional trade, a simulation of trade flows in the case of subregional droughts can be very interesting. It may be that intraregional transport can ensure consumption needs and contribute to an alleviation of price increases.

Figure 12.2—Regional deficit and surplus by province in three SADCC countries

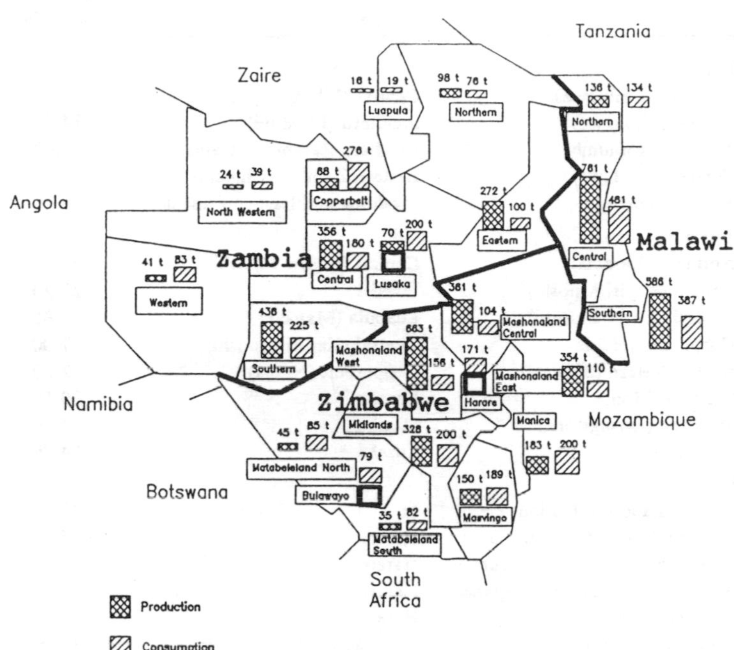

Sources: Data for Malawi are from U. Lele and S. W. Stone, *Population Pressure, the Environment, and Agricultural Intensification in Sub-Saharan Africa: Variations on the Boserup Hypothesis*, Managing Agricultural Development in Africa (MADIA) Study (Washington, D.C.: World Bank, 1989). Data for Zambia are from National Agricultural Marketing Board of Zambia, *Maize Storage in Zambia: A Feasibility Study* (Lusaka: NAMB, 1989). Data for Zimbabwe are from K. Muir and T. Takavarasha, "Pan-Territorial and Pan-Seasonal Pricing for Maize in Zimbabwe," in *Household and National Food Security in Southern Africa*, ed. G. D. Mudimu and R. H. Bernstein, 103-124 (Harare: University of Zimbabwe, 1989); Zimbabwe, Central Statistical Office, *Population Report* (Harare: CSO, 1985); and author's consultations with the World Food Programme and Southern African Development Coordination Conference (SADCC), Food Security Unit, Early Warning Department, 1989. The approach for deriving provincial levels is taken from SADCC, "Regional Food Security, Regional Food Reserve," a report prepared by Technosynesis for the Ministry of Agriculture, Harare, 1984.

Note: Weights are shown in thousands of metric tons.

Table 12.1—Transport routes and costs between provinces

From	To	US$ per Metric Ton
Malawi		
Northern (Chilumba)	Central (Lilongwe)	35.70
Central (Lilongwe)	Southern (Blantyre)	24.50
Northern (Chilumba)	Northern Zambia (Kasama)	78.00
Central (Lilongwe)	Eastern Zambia (Chipata)	12.70
Southern (Blantyre)	Mashonaland East (Nyamapanda)	23.90
Zambia		
Northern (Kasama)	Central (Kapiri Mposhi)	40.50
Central (Kapiri Mposhi)	Lusaka	22.20
Central (Kapiri Mposhi)	Luapula (Mansa)	29.80
Copperbelt (Ndola)	Central (Kapiri Mposhi)	17.00
North Western (Solwezi)	Copperbelt (Ndola)	12.50
Western (Mongu)	Lusaka	30.50
Southern (Livingstone)	Lusaka	54.70
Eastern (Chipata)	Lusaka	54.70
Zimbabwe		
Mashonaland West (Lions Den)	Lusaka	32.80
Mashonaland West (Lions Den)	Harare	15.80
Mashonaland Central (Bindura)	Harare	4.35
Mashonaland East (Nyamapanda)	Harare	21.40
Midlands (Gweru)	Harare	19.00
Midlands (Gweru)	Bulawayo	16.00
Matabeland North (Hwange)	Zambia Southern (Livingstone)	9.70
Matabeleland North (Hwange)	Bulawayo	32.90
Matabeleland South (Beit Bridge)	Bulawayo	27.90
Midlands (Gweru)	Masvingo (Masvingo)	10.35
Masvingo (Masvingo)	Manicaland (Mutare)	9.20
Manicaland (Mutare)	Harare	6.50
Bulawayo	South Africa (Johannesburg)	37.00
Manicaland (Mutare)	Mozambique Port (Beira)	27.00
South Africa		
Durban	Johannesburg	31.00

Source: Based on data from L. Berger, Southern Africa Regional Transportation Strategy Evaluator, Database Update (study prepared for U.S. Agency for International Development, Washington, D.C., 1986).

Detailed simulations are conducted in order to provide insight into what happens when shortfalls occur in certain subregions. The analysis draws upon information on the agroecological suitability of production in all provinces. Such information is provided by the Technosynesis report

(SADCC 1984), which has reviewed the relevant literature on agro-ecological conditions. Their approach of characterizing particular provinces as drought-prone by index numbers that are multiplied by the average amount of production in a province is drawn upon in this study. Figures measuring agroclimatic suitability can be found in Table 12.2.

In this approach, national variability is translated into subregional terms by applying coefficients of variation that have been corrected by the Technosynesis index. Since there are no reliable provincial data showing fluctuations in production, it seems reasonable to analyze risk in this manner. However, the assessment of production fluctuations on a provincial level should be improved as soon as better data is available.

Table 12.2—Agroclimatic suitability index

Country/Province	Technosynesis Index	Coefficient of Variation	Index Adjusted by Coefficient of Variation
Malawi		0.117	
Northern	0.97		0.86
Central	1.00		0.86
Southern	0.91		0.81
Zambia		0.176	
Northern	0.56		0.48
Luapula	0.50		0.42
Eastern	1.00		0.85
Central	0.92		0.79
Southern	0.95		0.81
Copperbelt	0.90		0.77
North Western	0.90		0.77
Western	0.93		0.79
Zimbabwe		0.295	
Mashonaland West	0.91		0.71
Mashonaland Central	0.91		0.71
Mashonaland East	0.91		0.71
Midlands	0.96		0.75
Matabeleland North	0.41		0.32
Matabeleland South	0.41		0.32
Manicaland	0.92		0.72
Masvingo	0.65		0.50

Source: Southern African Development Coordination Conference (SADCC), Regional Food Security Programme, "Regional Food Security, Regional Food Reserve" (main report and three annexes prepared by Technosynesis for the Ministry of Agriculture, Harare, 1984).

This Technosynesis index has been adjusted by time-series analysis for the individual countries, and the coefficients of variation have been calculated as a deviation of trends in production. Deviation from production trends can be seen in Figure 12.3. The highest coefficient of variation was calculated for Zimbabwe (29.5 percent), followed by Zambia (17.6 percent) and Malawi (11.7 percent).

An intercountry covariance analysis of deviations from production trends reveals little evidence that drought in Zimbabwe, for example, coincides with drought in Zambia. The correlation of production instability from 1960 to 1987 is shown in the table below. However, it could be of interest to investigate a region-wide drought such as occurred in 1983 and 1984, affecting mainly Zambia and Zimbabwe.

Country	Malawi	Zambia	Zimbabwe
Malawi	1.0000	0.4200	0.0019
Zambia	...	1.0000	0.1750
Zimbabwe	...	...	1.0000

Figure 12.3—Relative deviations of production from trend in Malawi, Zambia, and Zimbabwe, 1965-89

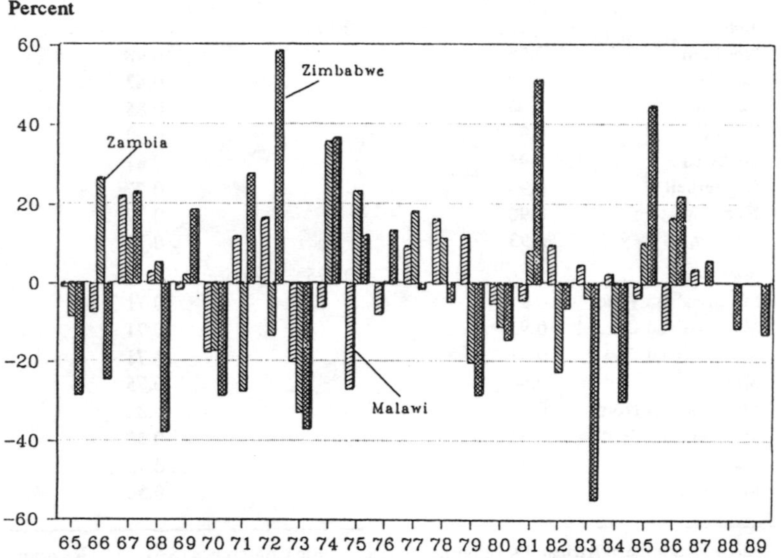

Source: Southern African Development Coordination Conference (SADCC), "Regional Food Security, Regional Food Reserve" (report prepared by Technosynesis for the Ministry of Agriculture, Harare, 1984).

Some arguments suggest that a drought affecting the whole region was more probable in the 1980s than in previous decades and may happen more often in the 1990s due to increased production in communal areas in Zimbabwe, which are more exposed to harvest fluctuations. However, these estimations of production instability may have to be adjusted after a trade liberalization.

Since a change in subregional prices could result in altered production patterns, a change in production instability might be expected. For example, a strong reduction of production in Mashonaland, which is comparatively suitable, would imply an increase in Zimbabwean production instability. On the other hand, if Malawi increases its proportion of regional production, reduced regional instability can be expected.

A drought that can be expected to occur with 16 percent probability (in other words, roughly every six years) was subjectively chosen for the simulation exercise. Simulating a whole set of different probabilities would provide policymakers with valuable information for planning different strategies. However, this study is designed to provide an example of trade flows that could prevail in the case of drought.

Integration of Stockpiling Activities in the Model

Since policymakers in the region are very concerned about drought, governments have introduced stockpiling. Huge stocks have been built up in order to avoid maize imports in the case of production shortfalls (that is, avoidance of foreign-currency expenditures). The economic rationale for these activities is based on the large differences between national f.o.b. and c.i.f. prices (Kingsbury 1989; Koester 1986). However, this argument is not always compelling, and regional trade, for example, could help reduce the need for stockpiling.

To address the problem of import aversion, the model runs link two periods of intraregional transport by introducing stockpiling activities. Stockpiling occurs as a transfer of overall surplus from year one, which is designated to be a normal year, to a second year. The second year is modeled as a drought year.

Several locations for stockpiling are chosen in the main surplus areas according to geographic conditions. The major reason for stockpiling in surplus areas is that producer prices are lower than consumer prices in these regions and transport costs can be saved. Hence, interest payments, which constitute a major part of stockpiling costs, are lower. It is believed at this stage of research that the advantages of stockpiling in surplus areas due to lower producer prices will not be outweighed by the immediate availability of stocks in consumption areas and lower construction costs in the mainly urbanized consumption areas.[1]

The model assumes that stockpiling facilities can be made available immediately at given costs; thus it does not take into consideration storage capacities that already exist in the region. This assumption is made to account for the possibility that the current location of stockpiling facilities is suboptimal but can be relaxed.

RESULTS

In this section, various model runs and their results are presented. Each discussion of individual measures begins with a description of the policy measures taken and ends with the presentation of their consequences as derived from model simulations. This is a subjective choice of possible policy options. However, it tries to reflect certain policy priorities in the southern African context. The presentation starts with total trade liberalization in a normal year.

Normal Year

To analyze changes in production, consumption, prices, and trade volumes due to policy changes, model runs are compared with a baseline run. A transport-cost-minimizing model run without adjustment in production and consumption serves as the baseline.

In Figure 12.4, which depicts this baseline run, it can be observed that trade between Malawi, Zambia, and Zimbabwe may emerge purely because of reduced transport costs. For example, it would be cheaper for the Zambian capital, Lusaka, to import from Zimbabwean Mashonaland West than from the more remote Southern Province of Zambia. The same applies to western deficit areas of Zimbabwe. Here one would expect imports from Zambia's Southern Province, even in the situation of no adjustment. Considerable trade volumes may prevail between subregions. All in all, transport costs are reduced by 12 percent compared with purely national procurements. This is a conservative estimate, since it employs optimized national transport and financial costs as a reference. In reality, savings might be higher due

---

[1] For more details on a discussion on the design of distribution and inventory systems see Boyd and Massy 1970, 509-539.

Figure 12.4—Regional trade using minimal transport costs

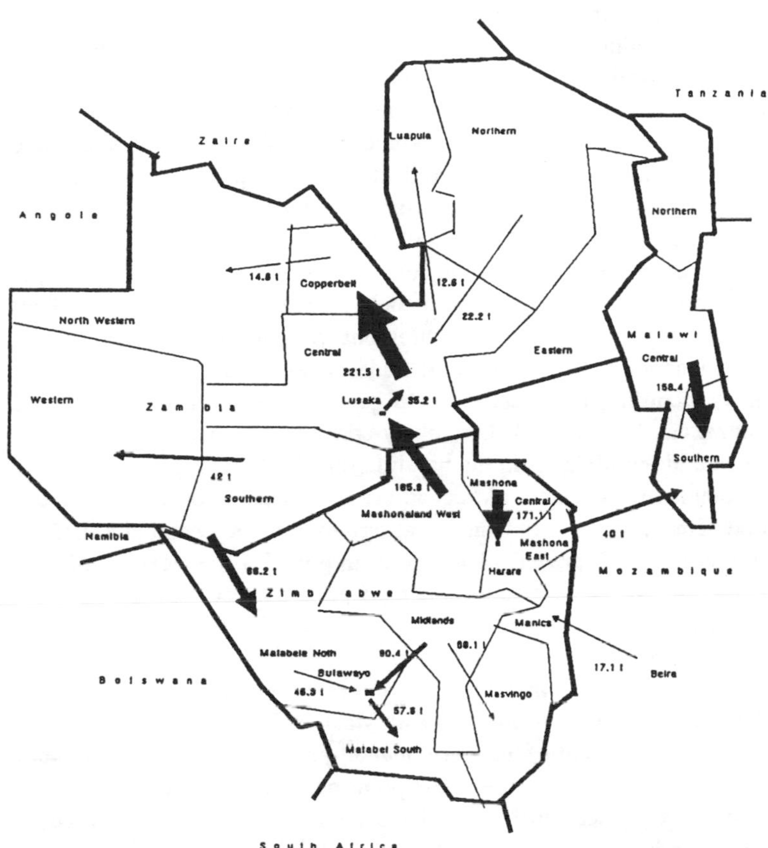

Sources: Data for Malawi are from U. Lele and S. W. Stone, *Population Pressure, the Environment, and Agricultural Intensification in Sub-Saharan Africa: Variations on the Boserup Hypothesis*, Managing Agricultural Development in Africa (MADIA) Study (Washington, D.C.: World Bank, 1989). Data for Zambia are from National Agricultural Marketing Board of Zambia, *Maize Storage in Zambia: A Feasibility Study* (Lusaka: NAMB, 1989). Data for Zimbabwe are from K. Muir and T. Takavarasha, "Pan-Territorial and Pan-Seasonal Pricing for Maize in Zimbabwe," in *Household and National Food Security in Southern Africa*, ed. G. D. Mudimu and R. H. Bernstein, 103-124 (Harare: University of Zimbabwe, 1989); Zimbabwe, Central Statistical Office, *Population Report* (Harare: CSO, 1985); and author's consultations with the World Food Programme and Southern African Development Coordination Conference (SADCC), Food Security Unit, Early Warning Department, 1989. The approach for deriving provincial levels is taken from SADCC, "Regional Food Security, Regional Food Reserve" (report prepared by Technosynesis for the Ministry of Agriculture, Harare, 1984).

Note: Weights are shown in thousands of metric tons.

to inefficiencies in the current transportation systems and higher opportunity costs for transport.

So far, adjustments in supply and demand permitted by trade have not been considered. Introducing trade and transport activities based on transport costs can be expected to result in adjustments in production and consumption. From the economic point of view, adjustments due to subregional price changes reflect marginal production costs and consumption preferences. Recursively, these changes may alter subregional procurement costs and imply changes in trade patterns.

Assuming that Beira is the sole source of imports at US$110 per metric ton, the next run of the model considers these feedback effects of trade on prices and thus on subregional production and consumption. In comparison with Figure 12.4, trade patterns do not change significantly according to Figure 12.5. In general, a slight drop in the trade volume can be expected. One likely change concerns the direction of maize trade in Malawi. In the baseline run, Malawi imports maize into its Southern Province. After adjusting prices and taking relative procurement costs into consideration, exports to Zambia's Eastern Province would prevail. Due to subregional price levels, Malawi becomes a low-price supplier to a high-price market (Eastern Province in Zambia) across its border. The resulting price levels are shown in Figure 12.6.

The lowest price in the region will prevail in the surplus subregions of Central Province in Malawi and Mashonaland East in Zimbabwe. Less than 40 percent of the world market price level will be obtainable. However, from a redistributional point of view, Malawian losses are small. Since prices in Malawi have already been 30 percent below world market prices in the situation before trade, an additional 10 percent would be only marginal. The opposite applies to Mashonaland, where prices would drop considerably due to liberalization.

In comparison, in a normal year, deficit subregions like the Copperbelt and Western provinces of Zambia will have prices 20-30 percent above world market levels. However, this is still of the magnitude of current pricing in Zambia. After trade liberalization, Harare, which is a subregion with a large population concentration, will enjoy a marginally depressed price level of 0-10 percent below the world market level. To a lesser extent, the same applies to the Zambian capital, Lusaka. Compared with prices that generally have been 20 percent above the world market level, market prices up to 10 percent below world market prices would be an improvement for consumers.

The Copperbelt Province is obviously critical. It is far away from surplus subregions and experiences significant increases in prices.

Figure 12.5—Trade after adjustment to price differences

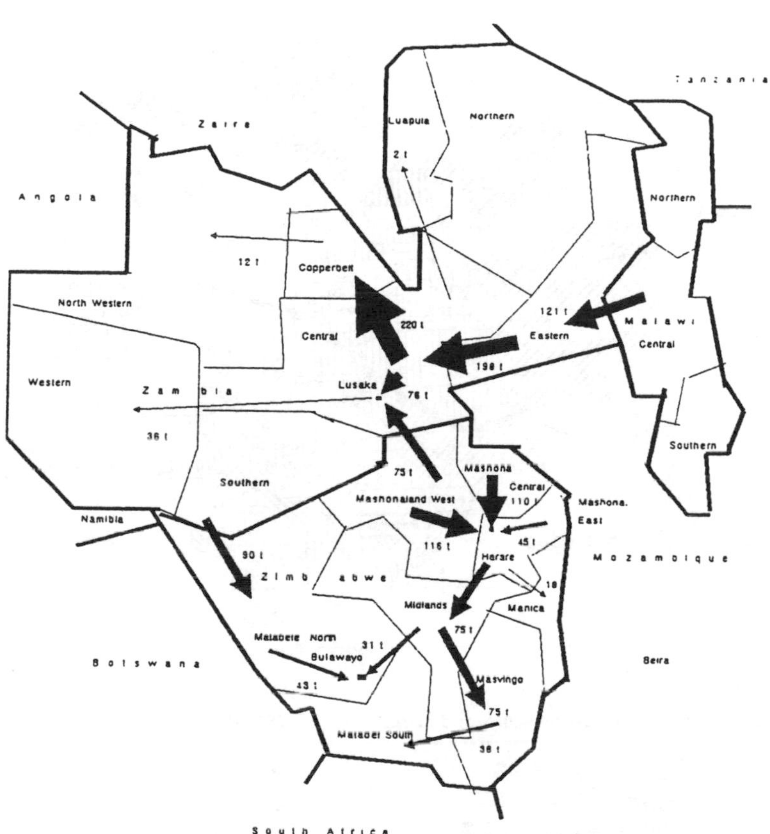

Sources: Data for Malawi are from U. Lele and S. W. Stone, *Population Pressure, the Environment, and Agricultural Intensification in Sub-Saharan Africa: Variations on the Boserup Hypothesis*, Managing Agricultural Development in Africa (MADIA) Study (Washington, D.C.: World Bank, 1989). Data for Zambia are from National Agricultural Marketing Board of Zambia, *Maize Storage in Zambia: A Feasibility Study* (Lusaka: NAMB, 1989). Data for Zimbabwe are from K. Muir and T. Takavarasha, "Pan-Territorial and Pan-Seasonal Pricing for Maize in Zimbabwe," in *Household and National Food Security in Southern Africa*, ed. G. D. Mudimu and R. H. Bernstein, 103-124 (Harare: University of Zimbabwe, 1989); Zimbabwe, Central Statistical Office, *Population Report* (Harare: CSO, 1985); and author's consultations with the World Food Programme and Southern African Development Coordination Conference (SADCC), Food Security Unit, Early Warning Department, 1989. The approach for deriving provincial levels is taken from SADCC, "Regional Food Security, Regional Food Reserve" (report prepared by Technosynesis for the Ministry of Agriculture, Harare, 1984).

Note: Weights are shown in thousands of metric tons.

Figure 12.6—Pattern of regional price differences

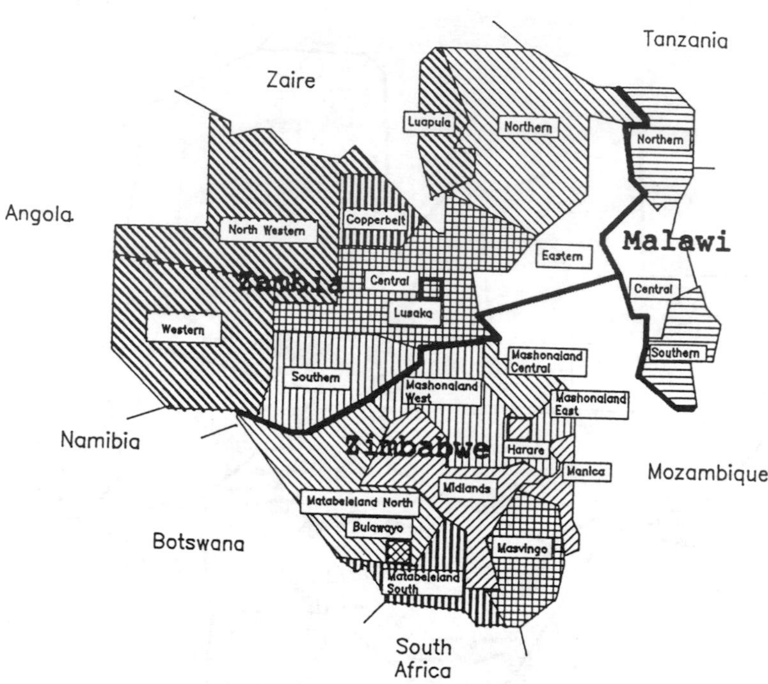

Price levels in comparison with world market price of US$110

(blank)	< − 40 %	(grid)	0 to 10 %
(horizontal lines)	− 40 to − 30 %	(cross-hatch)	10 to 20 %
(vertical lines)	− 30 to − 20 %	(dense vertical lines)	20 to 30 %
(diagonal lines)	− 20 to − 10 %	(dense diagonal lines)	30 to 40 %
(reverse diagonal lines)	− 10 to 0 %		

Sources: Data for Malawi are from U. Lele and S. W. Stone, *Population Pressure, the Environment, and Agricultural Intensification in Sub-Saharan Africa: Variations on the Boserup Hypothesis*, Managing Agricultural Development in Africa (MADIA) Study (Washington, D.C.: World Bank, 1989). Data for Zambia are from National Agricultural Marketing Board of Zambia, *Maize Storage in Zambia: A Feasibility Study* (Lusaka: NAMB, 1989). Data for Zimbabwe are from K. Muir and T. Takavarasha, "Pan-Territorial and Pan-Seasonal Pricing for Maize in Zimbabwe," in *Household and National Food Security in Southern Africa*, ed. G. D. Mudimu and R. H. Bernstein, 103-124 (Harare: University of Zimbabwe, 1989); Zimbabwe, Central Statistical Office, *Population Report* (Harare: CSO, 1985); and author's consultations with the World Food Programme and Southern African Development Coordination Conference (SADCC), Food Security Unit, Early Warning Department, 1989. The approach for deriving provincial levels is taken from SADCC, "Regional Food Security, Regional Food Reserve" (report prepared by Technosynesis for the Ministry of Agriculture, Harare, 1984).

However, this is a study of only three countries; Malawi, Zambia, and Zimbabwe. If, for example, integration with Zaire or Tanzania were to take place, the situation might change considerably. The potential for such changes can be seen in informal trade. Informal trade currently accounts for a considerable amount of shipment between Zambia's Copperbelt Province and Shaba Province in Zaire. Since all provinces bordering on Zambia are landlocked, world markets lose their potential for directly determining local prices.

To summarize, the price pattern will be determined by the relative surplus/deficit situation and distance from the port of Beira, which is assumed to be the sole gateway to international trade. However, this analysis is determined by initial price conditions. Hence, it only models a situation where partial liberalization of the maize economy takes place.

Distribution Consequences

As discussed above, the new pattern of prices will have distributional effects on consumers and producers. Consumers in peripheral subregions may suffer from higher prices. Large population groups in Lusaka and Harare would enjoy a relatively improved position. A different situation prevails on the supply side. In subregions where large-scale commercial farmers produce surpluses, price decreases will make the maize less favorable. This applies especially to Mashonaland. At the same time, these subregions have many alternatives in production that can compensate for losses in the production of maize, as reflected in the higher supply elasticity in these subregions.[2]

Looking at producer subgroups such as peasant farmers, surplus peasant farmers in remote subregions will be better off. Since their prices have been depressed so far, trade liberalization will give them incentives to develop local production and contribute to subregional food security. In contrast, net consumer households in the same subregion are negatively affected by higher prices. Their real income will be reduced.

Drought Year

In a simulation of policy options for improved regional food marketing, it is of interest to investigate how trade may alter sub-

---

[2] An extended model could provide evidence of relative price changes in crops competing for common resource use and deal more effectively with the question of relevant crop mix.

regional demand, trade, and prices in the case of a drought. The hypothesis can be stated that in comparison with autarky, trade could contribute to a relative improvement in the stability of local maize markets. Such a framework implies that prices vary with the availability of grain. This even takes place with the current situation of "controlled prices." Since governments are only partially successful in controlling prices, especially in rural areas, official prices are not always the relevant prices.

Model simulations presented in Figure 12.6 display possible price patterns in the case of a severe drought. As previously discussed, the model assumes the occurrence of a drought with a 16 percent probability.

The predicted production shortfalls are strongly dependent on the choice of probability. A reason for choosing that particular probability can be derived from the fact that it depicts the standard deviation of production. Intercountry correlation coefficients are not taken into consideration. Their introduction would require another dimension in the discussion. The drought simulated here occurs simultaneously in the three countries. In terms of the magnitude, the drought can be characterized by a regional shortfall of 840,000 tons or -21.6 percent of the normal production. This normal production has been drawn from the model of adjusted production, consumption, and prices. In summary, the regional shortfall means that Malawi would experience a national shortfall of 210,000 tons (-15.7 percent), Zambia of 160,000 tons (-15.8 percent), and Zimbabwe of 470,000 tons (-30.7 percent). Subregional values are according to the index of Table 12.2 and result in these national values.

Trade quantities depicted in Figure 12.7 prevail if trade liberalization allows for adjustments in the whole region. The simulation shows that Zimbabwe exports 160,000 tons to Lusaka, while Matabeleland imports 50,000 tons.

The corresponding price pattern can be observed in Figure 12.8. A national ad hoc policy alternative may be to close borders to maize trade in order to alleviate internal shortages. Such a situation, which models only national market liberalization, is displayed in Figure 12.9.

Clearly, price increases, especially in Zambia, are far more serious than in a situation with interregional trade (Figure 12.8). Since Mashonaland provinces contribute to the availability of grains in Zambia (Figure 12.7), price increases are more balanced in the whole region. Zimbabwean price increases will drop into the same category of 40-60 percent (Figure 12.9 compared with Figure 12.8) after trade liberalization. Malawi takes a more isolated position. Hence, a closure of borders for trade in maize by the Zimbabwean government would

Figure 12.7—Regional trade pattern in the case of a severe drought, border-crossing trade, second year

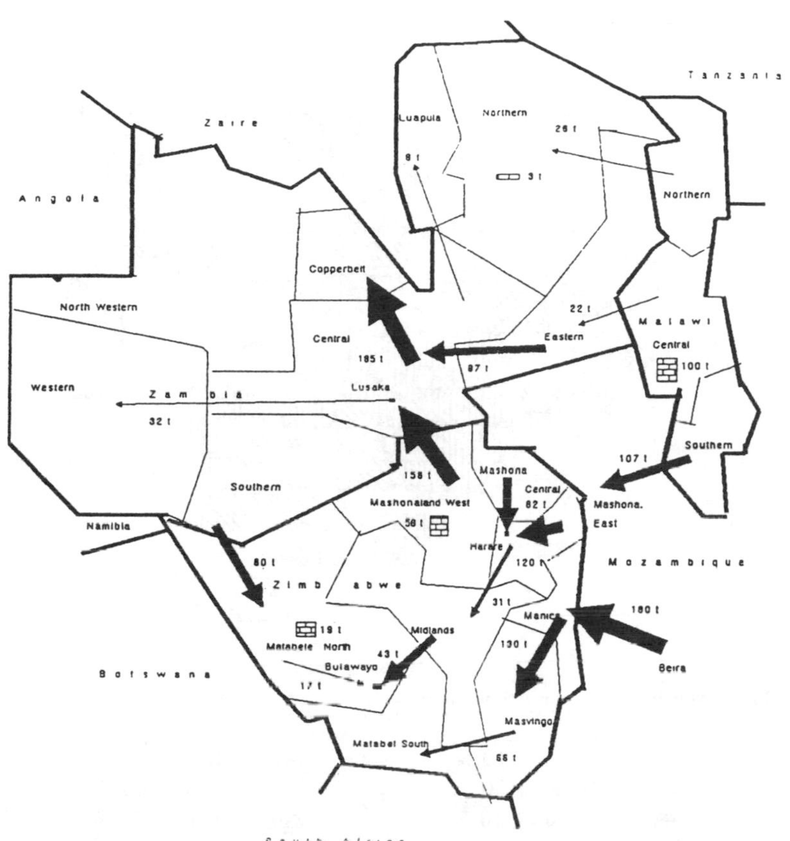

Sources: Data for Malawi are from U. Lele and S. W. Stone, *Population Pressure, the Environment, and Agricultural Intensification in Sub-Saharan Africa: Variations on the Boserup Hypothesis*, Managing Agricultural Development in Africa (MADIA) Study (Washington, D.C.: World Bank, 1989). Data for Zambia are from National Agricultural Marketing Board of Zambia, *Maize Storage in Zambia: A Feasibility Study* (Lusaka: NAMB, 1989). Data for Zimbabwe are from K. Muir and T. Takavarasha, "Pan-Territorial and Pan-Seasonal Pricing for Maize in Zimbabwe," in *Household and National Food Security in Southern Africa*, ed. G. D. Mudimu and R. H. Bernstein, 103-124 (Harare: University of Zimbabwe, 1989); Zimbabwe, Central Statistical Office, *Population Report* (Harare: CSO, 1985); and author's consultations with the World Food Programme and Southern African Development Coordination Conference (SADCC), Food Security Unit, Early Warning Department, 1989. The approach for deriving provincial levels is taken from SADCC, "Regional Food Security, Regional Food Reserve" (report prepared by Technosynesis for the Ministry of Agriculture, Harare, 1984).

Note: Weights are shown in thousands of metric tons.

Figure 12.8—Regional price pattern in the case of a severe drought, border-crossing trade

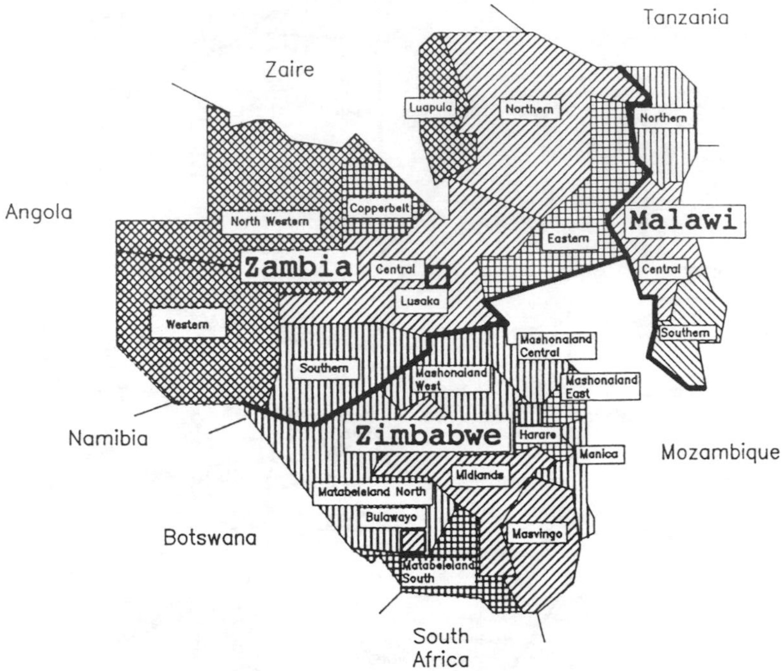

Price levels in comparison with world market price of US$110

- 20 to - 10 %
- 10 to 0 %
0 to 10 %
20 to 30 %
40 to 60 %
60 to 80 %
80 to 100 %

Sources: Data for Malawi are from U. Lele and S. W. Stone, *Population Pressure, the Environment, and Agricultural Intensification in Sub-Saharan Africa: Variations on the Boserup Hypothesis*, Managing Agricultural Development in Africa (MADIA) Study (Washington, D.C.: World Bank, 1989). Data for Zambia are from National Agricultural Marketing Board of Zambia, *Maize Storage in Zambia: A Feasibility Study* (Lusaka: NAMB, 1989). Data for Zimbabwe are from K. Muir and T. Takavarasha, "Pan-Territorial and Pan-Seasonal Pricing for Maize in Zimbabwe," in *Household and National Food Security in Southern Africa*, ed. G. D. Mudimu and R. H. Bernstein, 103-124 (Harare: University of Zimbabwe, 1989); Zimbabwe, Central Statistical Office, *Population Report* (Harare: CSO, 1985); and author's consultations with the World Food Programme and Southern African Development Coordination Conference (SADCC), Food Security Unit, Early Warning Department, 1989. The approach for deriving provincial levels is taken from SADCC, "Regional Food Security, Regional Food Reserve" (report prepared by Technosynesis for the Ministry of Agriculture, Harare, 1984).

Figure 12.9—Regional price pattern in the case of a severe drought, border-crossing trade not allowed

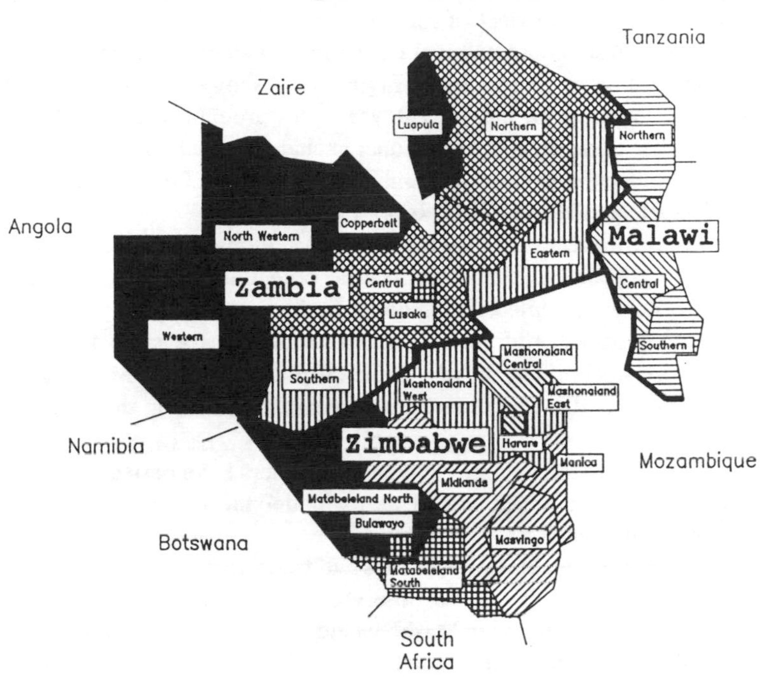

Price levels in comparison with world market price of US$110

20 to 30 %
30 to 40 %
40 to 60 %
60 to 80 %
80 to 100 %
> 100 %

Sources: Data for Malawi are from U. Lele and S. W. Stone, *Population Pressure, the Environment, and Agricultural Intensification in Sub-Saharan Africa: Variations on the Boserup Hypothesis*, Managing Agricultural Development in Africa (MADIA) Study (Washington, D.C.: World Bank, 1989). Data for Zambia are from National Agricultural Marketing Board of Zambia, *Maize Storage in Zambia: A Feasibility Study* (Lusaka: NAMB, 1989). Data for Zimbabwe are from K. Muir and T. Takavarasha, "Pan-Territorial and Pan-Seasonal Pricing for Maize in Zimbabwe," in *Household and National Food Security in Southern Africa*, ed. G. D. Mudimu and R. H. Bernstein, 103-124 (Harare: University of Zimbabwe, 1989); Zimbabwe, Central Statistical Office, *Population Report* (Harare: CSO, 1985); and author's consultations with the World Food Programme and Southern African Development Coordination Conference (SADCC), Food Security Unit, Early Warning Department, 1989. The approach for deriving provincial levels is taken from SADCC, "Regional Food Security, Regional Food Reserve" (report prepared by Technosynesis for the Ministry of Agriculture, Harare, 1984).

affect Zambian consumers considerably. The very costly alternative of importing grain from overseas has been neglected in order to demonstrate the particular effect of regional trade.

The corresponding national transport and stockpiling patterns can be seen in Figure 12.10. While Figure 12.7 shows an intensive trade between countries in the second year (the drought year), in Figure 12.10 other trade links (transnational exchanges) such as Mashonaland West (Zimbabwe) to Lusaka (Zambia) and Southern Province (Zambia) to Matabeleland North (Zimbabwe) are cut.

Zimbabwe seems to play the role of a regional stabilizer of prices, assisted by Malawi. Zimbabwe has the potential to export in drought years to Zambia, while keeping domestic increases relatively moderate at the same time. Additionally, stabilization by free trade is supplemented by stockpiling, mainly in the surplus regions of Mashonaland (see Figure 12.10). Stocks in drought years are lower in the free-trade scenario.

This discussion assumes that private traders have perfect foresight. An alternative approach would be to model the risk associated with imperfect information.

Comparing stockpiling activities and trade displayed in Figure 12.7 and Figure 12.10, it can be observed that imports from Beira can substitute for stockpiling in Mashonaland Central. These changes in the optimal activity pattern indicate that stockpiling enters into the solution only if no cheap, third-country grain resources are available. In other words, stockpiling is very sensitive to world market prices. This can be shown by a variation of stockpiling costs or world market prices.

To evaluate the current discussion, again it has to be emphasized that competitive imports from Beira are part of simulating a free-trade situation. Model runs with no external trade via Beira as a particular policy option, however, would not substantially change the internal trade pattern in the second period. This has been checked by various model runs not displayed. Since more stocks must be made available as a substitute for external trade, it is procurements for stocks in the first period that are mainly affected.

External trade can be financed by exports in surplus years. To evaluate these export chances, a model run simulating a bumper crop could be helpful. In theory, a bumper harvest has the same probability as a drought. In some years Zimbabwe and Zambia do not know where to dispose of their surpluses at reasonable prices. In other words, third-country exports in bumper years could finance imports in drought years. Perhaps, partly because of the difference between import and export parity prices, less finance could be made available. Unfortunately,

Figure 12.10—Regional trade pattern in the case of a severe drought, without border-crossing trade, second year

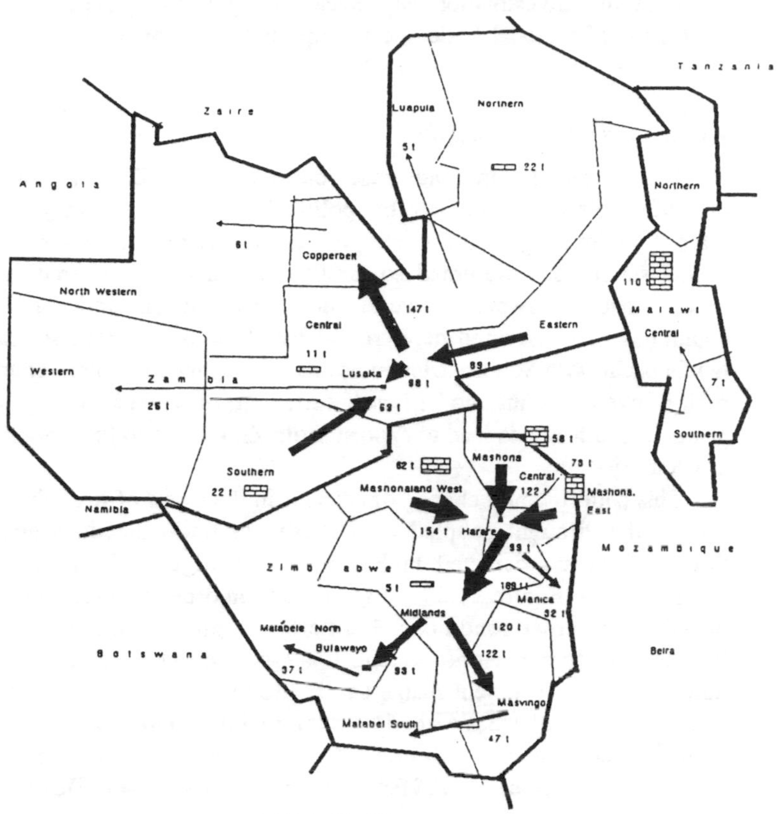

Sources: Data for Malawi are from U. Lele and S. W. Stone, *Population Pressure, the Environment, and Agricultural Intensification in Sub-Saharan Africa: Variations on the Boserup Hypothesis*, Managing Agricultural Development in Africa (MADIA) Study (Washington, D.C.: World Bank, 1989). Data for Zambia are from National Agricultural Marketing Board of Zambia, *Maize Storage in Zambia: A Feasibility Study* (Lusaka: NAMB, 1989). Data for Zimbabwe are from K. Muir and T. Takavarasha, "Pan-Territorial and Pan-Seasonal Pricing for Maize in Zimbabwe," in *Household and National Food Security in Southern Africa*, ed. G. D. Mudimu and R. H. Bernstein, 103-124 (Harare: University of Zimbabwe, 1989); Zimbabwe, Central Statistical Office, *Population Report* (Harare: CSO, 1985); and author's consultations with the World Food Programme and Southern African Development Coordination Conference (SADCC), Food Security Unit, Early Warning Department, 1989. The approach for deriving provincial levels is taken from SADCC, "Regional Food Security, Regional Food Reserve" (report prepared by Technosynesis for the Ministry of Agriculture, Harare, 1984).

Note: Weights are shown in thousands of metric tons.

a full discussion of these or related questions would require a multiperiod stochastic simulation model that exceeds the capacity of this study. Further investigation may reveal some interesting aspects of the interaction of regional trade and multiperiod behavior of a liberalized maize economy.

Variation in Residual Supply

Since external trade and trade routes in the SADCC region are presently determined in part by political constraints, it may be of interest to investigate the consequences of lifting such constraints. Model simulations presented so far have not investigated trade with South Africa. However, several model simulations under normal conditions, not discussed here, reveal that little or no trade in maize would occur with South Africa in normal year. Since the three-country region covered by the model is self-sufficient, only unexpectedly high prices in South Africa lead to exports from Zimbabwe to that neighboring country.

This pattern could change drastically in the event of drought. The results of a drought coupled with a drop of trade sanctions against South Africa are depicted in Figure 12.11. Figure 12.11 must be compared with Figure 12.8 (or Figure 12.9) in order to understand the underlying internal conditions. Furthermore, procurement conditions from South Africa must be evident. It is assumed that South Africa has no drought and sufficient maize can be made available at a comparatively low price of US$74 at silos in the region around Johannesburg. This price corresponds with internal freight rates in South Africa and an f.o.b. import price of US$105 at the harbor of Durban. This setting reflects opportunity costs for international exports in South Africa. Given a transport rate of US$37 from Johannesburg to Bulawayo by rail, this means a price of US$111 in Bulawayo. Accordingly, any other price change in comparison with Figure 12.8 will be less than in the reference situation of internal trade with Beira as the sole source of imports.

In total, a volume of 420,000 tons of maize is shipped mainly into Zimbabwean drought subregions, thus enabling Zimbabwean shipments to Zambia. Again, a more isolated position is taken by Malawi, which shows no price adjustments to the new trade opportunities. A variation in the South Africa supply price may change these results. For example, internal producer prices of South Africa have been reported as US$100 in 1989 (see Chapter 14). As a consequence of such prices, exports to Zimbabwe would be 137,000 tons, and Beira would be the

Figure 12.11—Regional price pattern in the case of a severe drought, border-crossing trade including South Africa

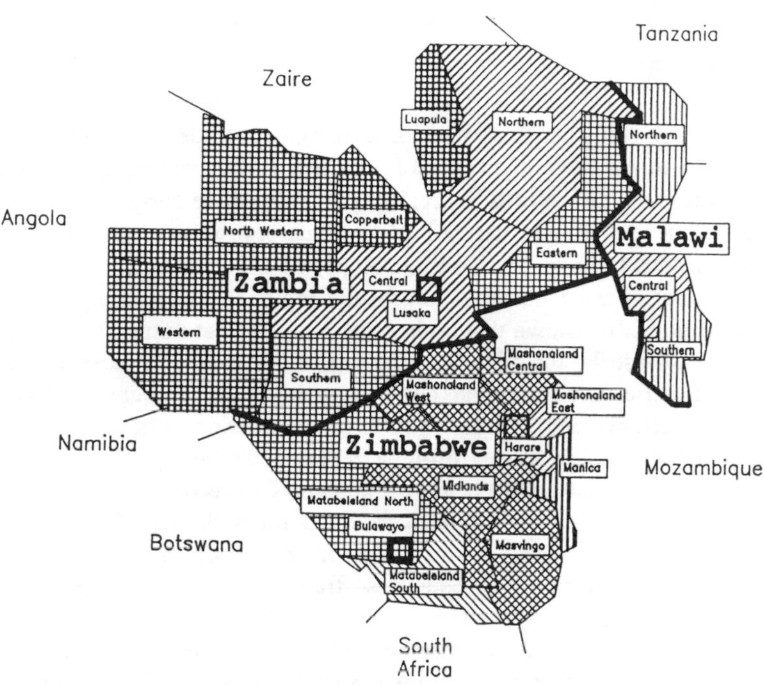

Price levels in comparison with world market price of US$110

- 20 to - 10 %
- 10 to 0 %
0 to 10 %
10 to 20 %
20 to 30 %
40 to 60 %
60 to 80 %

Sources: Data for Malawi are from U. Lele and S. W. Stone, *Population Pressure, the Environment, and Agricultural Intensification in Sub-Saharan Africa: Variations on the Boserup Hypothesis*, Managing Agricultural Development in Africa (MADIA) Study (Washington, D.C.: World Bank, 1989). Data for Zambia are from National Agricultural Marketing Board of Zambia, *Maize Storage in Zambia: A Feasibility Study* (Lusaka: NAMB, 1989). Data for Zimbabwe are from K. Muir and T. Takavarasha, "Pan-Territorial and Pan-Seasonal Pricing for Maize in Zimbabwe," in *Household and National Food Security in Southern Africa*, ed. G. D. Mudimu and R. H. Bernstein, 103-124 (Harare: University of Zimbabwe, 1989); Zimbabwe, Central Statistical Office, *Population Report* (Harare: CSO, 1985); and author's consultations with the World Food Programme and Southern African Development Coordination Conference (SADCC), Food Security Unit, Early Warning Department, 1989. The approach for deriving provincial levels is taken from SADCC, "Regional Food Security, Regional Food Reserve" (report prepared by Technosynesis for the Ministry of Agriculture, Harare, 1984).

source of a further 138,000 tons (according to model runs not displayed). These variations primarily affect Zimbabwe, where price levels would increase 20-30 percent (Figure 12.11).

SUMMARY

In this study, the potential for intraregional trade and subregional adjustment has been investigated for three SADCC countries: Malawi, Zambia, and Zimbabwe. Subregional adjustment in supply, demand, and trade patterns is determined by changes in subregional price patterns due to trade liberalization. Applying a spatial partial equilibrium model of trade liberalization for the regional staple food product, maize, it can be shown that trade contributes to cost-minimal procurement and distribution of food in the region. In spite of a current level of self-sufficiency of around 100 percent in the three countries, trade with a neighboring country would be a superior policy option to national marketing alone. From a regional welfare point of view, neighboring countries are better off opening their economies for maize trade. Assuming regional welfare maximization without preferences for any of the participating groups— consumers, producers, and governments in each of the three countries—trade will be part of a regional strategy of efficient marketing.

However, introducing intraregional trade will lead to subregional price differences that reflect transport costs. This in turn will lead to adjustments in trade patterns. Due to the model design of supply and demand responses allowed within SADCC and with South Africa, consumers and producers adjust to price changes simultaneously. As expected, the surplus Zimbabwean Mashonaland provinces and the regionally remote Malawian provinces will have the lowest prices, while deficit subregions like the Copperbelt in Zambia and southwestern Zimbabwe will be confronted with considerably higher prices. Hence, from a distributional point of view, producers in surplus subregions and consumers in remote deficit regions will be worse off. Interestingly, consumers in the capitals Lusaka and Harare will enjoy a moderate price drop compared with the current situation of guaranteed panterritorial prices. Since these regions are not too far away from surplus subregions or third-country import-opportunity locations such as Beira, or both, price levels will have a tendency to fall below world market prices.

Furthermore, the effects of trade in the case of a severe drought in the whole region have been investigated. It has been shown that intraregional trade can soften price increases to some extent, especially in Zambia, and thus can contribute to an alleviation of hardships from drought. In the event of drastic price increases, imports from neighboring countries can alleviate some food shortages. This stabilization via neighboring countries leads to only small price increases in the country of origin—Zimbabwe in the present case. However, only a specified situation has been investigated, relating to past data and probability assessments of drought occurrence. Hence, this particular approach serves primarily as an illustration of the effects of alternative policies.

Additionally, third-country imports may still play an important role. It has been shown that Zimbabwe should import from Beira and ship its own grain further north to Zambia. These results are according to theoretical expectations about trade as an iterative exercise using price differences for cost-minimal transports of food staples.

REFERENCES

Bale, M. D., and E. Lutz. 1981. Price distortions in agriculture and their effects: An international comparison. *American Journal of Agricultural Economics* 63 (1): 8-22.

Berger, L. 1986. Southern Africa regional transportation strategy evaluator. Data base update. Study prepared for U.S. Agency for International Development, Washington, D.C.

Boyd, H. W., and W. Massy. 1970. *Marketing management*. The Harbrace Series in Business and Economics. New York: Harcourt, Brace, Jovanovich.

Buccola, S. T., and C. Sukume. 1987. Welfare-optimal agricultural policy in less developed economies. Draft of Department of Agricultural and Resource Economics, Corvallis, Oregon.

Hazell, P. B. R., and R. D. Norton. 1986. Mathematical programming for economic analysis in agriculture. New York: Macmillan.

Kandoole, B., B. Kaluwa, and S. Buccola. 1989. Market liberalization and food security in Malawi. In *Southern Africa: Food security policy options*, ed. M. Rukuni and R. Bernstein, 101-118. Proceedings of the Third Annual Conference of the University of Zimbabwe/Michigan State University Food Security Research Project, 1-5 November 1987, Harare. Harare: University of Zimbabwe.

Kingsbury, D. 1989. Agricultural pricing policy and trade in several SADCC countries: Preliminary results. In *Household and national food security in southern Africa*, ed. G. D. Mudimu and R. H. Bernstein, 259-276. Proceedings of the Fourth Annual Conference of the University of Zimbabwe/Michigan State University Food Security Research Project, 31 October - 3 November 1988, Harare. Harare: University of Zimbabwe.

Koester, U. 1986. *Regional cooperation to improve food security in southern and eastern African countries*. Research Report 53. Washington, D.C.: International Food Policy Research Institute.

_____. 1990. Agricultural trade among Malawi, Tanzania, Zambia, and Zimbabwe: Competitiveness and potential. Paper prepared at the request of the World Bank by the International Food Policy Research Institute, Washington, D.C.

Lele, U., and S. W. Stone. 1989. *Population pressure, the environment, and agricultural intensification in Sub-Saharan Africa: Variations on the Boserup hypothesis*. Managing Agricultural Development in Africa (MADIA) Study. Washington, D.C.: World Bank.

Muir, K., and T. Takavarasha. 1989. Pan-territorial and pan-seasonal pricing for maize in Zimbabwe. In *Household and national food security in southern Africa*, ed. G. D. Mudimu and R. H. Bernstein, 103-124. Proceedings of the Fourth Annual Conference of the University of Zimbabwe/Michigan State University Food Security Research Project, 31 October - 3 November 1988, Harare. Harare: University of Zimbabwe.

National Agricultural Marketing Board of Zambia. 1989. *Maize storage in Zambia: A feasibility study*. Lusaka: NAMB.

SADCC (Southern African Development Coordination Conference). 1984. Regional food security, regional food reserve. Main report and three annexes prepared by Technosynesis for the Ministry of Agriculture, Harare.

_____. 1982. *A Regional Inventory of Agricultural Resource Base: Food Security Feasibility Study*. Vol.3, *Country Reports*. Harare: SADCC.

Takayama, T., and G. G. Judge. 1971. *Spatial and temporal price and allocation models*. Amsterdam.

Valdés, A. 1990. Preliminary estimates of exchange rate misalignment for Malawi, Zambia, and Zimbabwe. Paper presented at the IFPRI/SADCC Policy Workshop on Trade in Agricultural Products among SADCC countries, 27-28 February, Harare.

Zimbabwe. 1987. A report on SADCC Regional Food Security Project 1: A general technical assistance programme designed to achieve co-ordination and co-operation on all agrarian issues. Harare: Government Printer.

Zimbabwe, Central Statistical Office. 1985. *Population report*. Harare: CSO.

Part IV
Potential for Intraregional Trade

13
Policies for Promoting Agricultural Trade Between Malawi, Zambia, and Zimbabwe

Ulrich Koester

Malawi, Zambia, and Zimbabwe are in the process of pursuing structural reforms and liberalization policies. An expected outcome is an expansion of agricultural trade. The author has given evidence of a trade potential in Chapter 3. The present chapter deals more specifically with the political issues of trade liberalization. It starts with a brief summary of the rationale for policy reforms in the countries under consideration. It is argued that policy reform has to be seen in the light of the past bad economic performance of the countries. A policy reform, which may cause some hardship for some groups in the society, may be considered as a necessary process only if the population has been convinced that past policies cannot be pursued in the future. Thus, the section on the rationale for policy reforms is supposed to set the stage.

In a following section, specific emphasis will be laid on the role of the government in the reform process. It is a widely held belief, but misleading, that liberalization just implies less governmental activity. It is argued here that an active government is needed to make the liberalization process successful.

It is certainly true that any liberalization policy should be designed according to the specifics of the country under consideration. Nevertheless, some general guidelines for the case of the three countries under investigation here are provided in a later section, followed by a discussion of alternative liberalization and trade-promotion policies that can be instituted by joint actions by a group of countries or unilaterally by the individual countries.

THE RATIONALE FOR POLICY REFORMS

Policymakers may be reluctant to change present policies because to do so suggests that these policies were inadequate in the past and are not appropriate for the future. Apart from the adequacy of past policies, the challenges that face these countries are a sound rationale for genuine policy changes.

First, given their high external indebtedness and their limited access to foreign borrowing, the individual countries have to adjust their internal consumption level closer to their production capacity.

Second, past policies were originally instituted to control the economy. However, this approach has led to a growing public sector and a growing budget deficit. As private agents have found numerous ways of avoiding the controlled sectors and moved to the informal sectors (Lal 1987), governments have lost control and the tax base has declined, resulting in lower tax revenues over time and an increase in the budget deficit. Policy reforms could provide the means for governments to regain control over the economy.

Third, in most African countries, one objective of past policies was to address the food-security issue. It was felt that food security could not be left to the vagaries of the markets. However, on the one hand, as noted earlier, governments have not been very successful in enforcing regulations on domestic markets and have incurred increasing costs in the process. On the other hand, present knowledge on the functioning of markets and the determinants of food security supports the argument that markets can be used efficiently to achieve food security.

Fourth, the three countries concerned are in a better position now than in the past to provide the institutional support necessary to a market system. Such governmental institutions include well-defined property rights, a legal system, and a marketing-information system. In addition, markets can function better at present because communication costs, one of the main determinants of transaction costs, have been declining over time.

Fifth, countries are well advised to take into account the experience of other countries to identify the key issues and to avoid in the future the mistakes made in the past. There is a widely accepted view that too strongly regulated economies are doomed to fail economically. The breakdown of the eastern European communities illustrates the case in point. Many countries (including Malawi and Zambia) have implemented structural adjustment and liberalization policies; others are just starting (Zimbabwe, for example). The experiences have been mixed; Zambia just suspended a major part of its adjustment policy. Hence, the timing may be good for a discussion of liberalization issues at a regional conference.

The liberalization of the internal food economy is expected to give rise to additional trade among the three countries. However, it would not be advisable to implement a liberalization policy in one sector and ignore the other sectors of the economy. As will be argued later, one sector may lead the liberalization process, but it must be part of an economywide liberalization policy. In this context, a discussion of general policy reform issues will be presented before related issues specific to agriculture.

THE ROLE OF THE GOVERNMENT

There is ample evidence from the experience of other countries that the introduction of a market economy cannot be achieved solely by allowing market forces to work and by withdrawing all public activities. Rather, the move to a market economy demands that the government play an active role in providing support to the economic activities of private agents.

Securing the Political Acceptance of the Policy Reform

Judging from the experience of other countries, the failure or limited success of many policy reforms is not due to the technical incompetence of the governments, but rather to the political opposition that resists changes or makes the reform less effective by being uncooperative. How can a government provide a political environment that supports the objectives of the policy reform?

Nelson (1984, 985) argues that the factors determining political sustainment are the strength of political leaders' commitment to the program, the government's capacity to implement the program and manage political responses, and the political response the program evokes from influential groups.

Any policy change redirects income flows; hence, previously favored groups may lose out in favor of other groups in the society. In most cases, the losers from the reform are easily identified because they belong to small, urbanized, and better-organized groups with political influence. On the other hand, the prospective gainers from the policy change are usually from rural areas, in some cases distant from the political centers, and numerous, hence less organized. More to the point, the gains may materialize later than the losses, and the society may incur additional costs associated with the adjustment process. Thus there is a permanent need for the government to educate the population on the proposed policy reform. Such activities may be crucial for the survival of the government or for the self-interest of the policymakers who pursue the policies. What strategy should the government use to promote these policy reforms?

Some governments have tried what may be called the "scapegoat" approach. The pressure for policy reforms is in most cases exerted from outside the country, notably from the World Bank and the International Monetary Fund. Policymakers have attempted to blame these two organizations for the hardship that has followed the implementation of the policy change. The intention is to redirect the opposition to an external scapegoat. There is the inherent danger in this approach that some internal opposition groups might accuse the government of succumbing to foreign rule, which of course cannot serve the interest of the country, and thus force the government to change its policy course. The scapegoat approach is not in most cases a viable strategy. If the society at large is to be convinced of the necessity for and benefits of the policy change, the policymakers must unequivocally support the policy. A viable strategy should start by conveying to the public the strong commitment of the government to the policy reform.

Private agents have to be convinced that the reform process will be pursued. This is important because the policy requires major and rapid adjustments on the part of the private sector. Therefore, the government has the responsibility to signal unequivocally in which direction it is going to steer the economy, independently of whether the new policy is phased in gradually or introduced in one stroke. Of course, this does not mean that the government should make specific statements on the future state of the economy. It would be too risky to pretend to know too much. What is needed is clarification of the principles that guide governmental activities; the government should adhere to more rules and make fewer discretionary decisions.

The success of the reform depends on broad public support, hence the government would be well advised to discuss policy problems openly and above all to inform the public, on an ongoing basis, on the indicators of success. This will ensure the support of those who gain from the policy change and the development of a political constituency for reform (Hawkins 1991, 847).

At the same time, the government must be prepared to face the opposition arising from those groups that are adversely affected by the policy reforms. After the reform the old constituents may lose their political power and even a source of supplemental (sometimes illegal) revenue. If, for example, licenses are not needed anymore, the individuals who were formerly in charge of allocating licenses will lose power and access to an additional income. Likewise, with the reduction of governmental bureaucracy, many bureaucrats who were needed to implement the old policies may become unemployed. A viable reform process requires that the government deal with these issues in some compensatory manner. One way would be for the government to gradually decrease the number of bureaucrats, allowing time for the formal sector to gain momentum and absorb the freed labor force. Moreover, there may be a need to compensate civil servants for their loss of employment. This may slow the reduction of public expenditures, but it may be a rational means to secure a smooth policy-reform process.

Most likely, there is a group that will not easily adjust to the change in economic conditions and whose food security may be threatened. Indeed, if a consequence of the policy reform is impaired food security for some parts of the society, and if the government has no program to deal with it, it may cause the government to give up the reform. The following section proposes some guidelines in addressing this group's food-security concerns.

Taking Care of the Poor

The greatest concern of the government should be the impact that the reform will have on the poor. This group is the least prepared to face the adverse effects of policy reforms. Some governmental intervention may be needed. However, these activities must be country-specific because their effectiveness depends on the government's ability to implement social policies and the country-specific set of alternatives.

First, these activities should target the group whose welfare the government is trying to improve. Therefore, this group must be identified. For the countries under consideration, most of the poor live in rural areas. Hence, it is important for the government to determine how liberalization and structural adjustment policies will affect the rural sector.

It can be expected that one outcome of liberalization will be higher agricultural producer prices. On the one hand, higher producer prices translate into higher income for rural areas and generally improve welfare. On the other hand, there is a widely held view that grain-deficit households in rural areas may become more food-insecure from a rise in grain prices. However, this view has been challenged by recent research findings based on information on alternative sources of revenues for grain-deficit households.

Sahn and Sarris (1991) identify in their study three categories of income: earnings from agricultural activities, earnings from nonagricultural activities, and nonearned income. They found that, for example, low-income rural smallholders in Malawi's Southern Province generated 51 percent of their total income from agriculture; nonearned income (remittance) came to 36 percent of total income in 1989.

A survey from the World Bank (1990) supports these findings. Most of the poor live in rural areas, and the poorer they are, the higher the share of off-farm income in total income. It should be noted that one determinant of being poor is the lack of employment opportunities. According to the World Bank study, the core poor—the poorest 20 percent—work only 532 days per household per year, while the nonpoor work 762 days per household per year. Stanning (1989) found similar results for Zimbabwe. She conducted a survey in three low-income rural areas: Hurungwe, Bushu, and Binga. The share of net farm income in full income was 34.9 percent, 23.2 percent, and 30.9 percent for the households in the respective regions. These data reveal that increased grain prices alone may not result in increased food insecurity for grain-deficit households. Several other factors have to be investigated.

- How does the policy reform affect income received from sources other than grain production? Will the overall development of the rural areas provide better off-farm employment opportunities for grain-deficit farmers? Information on alternative sources of income for grain-deficit households is needed.
- It is now well documented that parallel markets are active in all three countries and that prices on these markets differ significantly from those on the official market. With the deregulation of the agricultural sector, grain prices may be expected to be higher than past official prices but not necessarily higher than prices from parallel markets. More important, allowing market forces to work and improving the functioning of markets will lead to greater price stability than now exists on the parallel markets.
- Furthermore, higher grain-producer prices do not necessarily imply higher consumer prices. But liberalizing the markets will most likely minimize the margin between consumer and producer prices. This expectation is based on a recent development on the Zimbabwean maize market and a new research finding.

    The Zimbabwean government started in 1990 to set producer prices for yellow maize 15 percent lower than producer prices for white maize. There is, of course, a sound rationale for such a price differentiation. However, at the same time, the government allowed producers of yellow maize to sell their produce directly to livestock farmers in their region. Thus the users of yellow maize could buy their feed either from the Grain Marketing Board (GMB) or directly from the grain producer. However, as the selling price of the GMB was much higher than their buying-in price, users of grain were better off buying directly from producers of yellow maize. Producers were better off selling directly to users because they could get a price that was not only higher than the GMB buying price for yellow maize but also higher than the GMB buying price for white maize. This example clearly indicates that the margin between producer prices and consumer prices will change significantly if private trading is allowed, and higher producer prices will not necessarily mean higher consumer prices.

A study on informal grain marketing in Zimbabwe found that the controlled price of commercial maize meal was between 10 and 80 percent higher per kilogram than the price for maize obtained and milled through informal channels (Chisvo et al. 1991). This price differential exists in spite of the high transaction costs on the informal markets and the higher quality of locally processed meal. Liberalizing markets may thus even improve the food-security situation in rural areas by encouraging markets of locally processed grains.

The observations presented above indicate that food security might not be negatively affected as the opponents of the policy reform argue. Nevertheless, the government must be prepared to survey the effects of the adjustment process and to design special programs for the poor. Fortunately, the governments can draw on the experience of other countries or even on their own experience. The lesson to be drawn from Zambia's policies to protect the poor is not very encouraging. Policies that depress prices of products that are easily traded across country borders seem to be counterproductive. These policies encourage illegal transborder trade that may contribute to more internal food insecurity and to a general loss in the welfare of the country.

In some cases (for example, in Zimbabwe) it might be possible to subsidize the prices of products that are mainly consumed by poor households and are not extensively traded across borders. For instance, a subsidy program could target a product like locally processed maize meal, which is rougher than the commercially processed kind but is preferred by poor households (Chisvo et al. 1991). The Zambian experience with the coupon system should also be investigated.

Securing the Institutional Framework

Governments that control the economy tend to violate a genuine division of labor between the government and the private sector. The task of the government can best be characterized by a quote from Keynes, "The important thing for government is not to do things which individuals are doing already, and to do them a little better or a little worse, but to do things which at present are not done at all" (World Bank 1991, 128). Hence, the government should withdraw from private economic activities, but instead should provide the institutional framework that supports the functioning of a market economy. This includes institutions that maintain law and order, invest in essential infrastructure, and secure property rights, contracts, and norms of conduct.

The experience of the European Community (EC) supports the view that the institutional framework that is instituted at the very beginning is crucial for the performance of any integration scheme. Hence, some guidelines are outlined as follows:

1. The negotiations should focus on general rules rather than specific regulations. As much as possible, instituting a decisionmaking framework that demands frequent discretionary decisions should be avoided. The problems on the agricultural markets in the EC would have never arisen but for the specific institutional framework that had been set up at the outset and is very difficult to change. Any change would most likely harm the interests of some member countries and therefore they would not agree on the change.

2. One of the most general rules accepted concerns the nature and magnitude of the common external protection. Only tariffs should be allowed; the rate of tariffs should be uniform across all products. Furthermore, there should be an agreed time frame for tariff reductions.

3. Countries should agree to abolish all nontariff barriers to intraregional trade. The general rule should be that all products that are on the market in one member country should be allowed on other national markets as well. Countries may only demand that the origin and product quality has to be declared on the label of the product.

4. Of special importance is the financial system that will be chosen to finance common policies and institutions. The EC accepted the principle of financial solidarity. The main EC institution, the European Commission, has access to funds to finance common policies. The rationale behind this is that common policies represent the political will of all member countries, hence these policies have to be financed out of a common fund. No doubt the rationale is convincing, but the implementation can have significant negative side effects. These effects may materialize if individual countries have to contribute to common

finances independently of national costs and benefits from the common activities and if countries are able to increase their net benefit by autonomous national activities. Thus, the principle of financial solidarity can give rise to externalization; individual member countries try to improve their national welfare at the cost of the partner countries. The Common Agricultural Policy of the EC suffers from this fact. The lesson to be drawn is that the design of the financial system should not allow any member country to externalize the costs of autonomous national decisions.

5. Countries have to agree on some harmonization of national policies. It is clear that they have to abstain from any system of administered prices. But it might be less obvious that they have also to harmonize the system of indirect taxes, notably value-added taxes. Otherwise, they cannot abolish all controls of cross-border trade. Moreover, at least some coordination of monetary policies is needed to avoid overvaluations of currencies or frequent exchange-rate realignments.

6. It is well known from international institutions such as the United Nations Organizations and the General Agreement on Tariffs and Trade (GATT) that it is much easier to find an agreement than to enforce it. Hence, it is necessary to set up a Common Court of Justice from the very beginning. The task of the court would be to decide on cases in which countries possibly do not adhere to the rules of the agreement. Each individual company or person in the newly created community should be allowed to bring a case to the court.

7. In order to enforce the decisions of the court it is advisable to set up a Common Fund. Member countries should be prepared to deposit a certain amount of foreign exchange that could bear interest on their account. If the court rules against a country, those who have been damaged should be compensated by payments from the Common Fund.

8. As stated above, a regional integration scheme will be successful only if it is used as a step to worldwide integration. Hence, regional integration should be a part of a liberalization policy. Certainly, liberalization demands significant adjustments in the internal economies, and there will be gainers and losers. Whether the necessary adjustment will actually take place depends very much on the credibility of the new policies. Thus the governments would be well advised to support institutions that support the credibility of the new policies. One of these institutions could be an independent Council of Economists. The main tasks of the council would be to survey and assess policies on the community and national levels and thus inform the public on whether the policymakers adhere to the agreed rules. The council would thus

serve as a transparency agency. Actually, the council could help the governments to argue against those groups in the economies who lose by the liberalization policies and who demand special protection.

GENERAL GUIDELINES FOR PHASING-IN AGRICULTURAL TRADE PROMOTION POLICIES

All three countries have carried out export-promotion policies in the past. However, these policies may have been harmful for trade expansion because they focused on only one side of the trade balance. Trade is supposed to be an exchange of commodities, and trade promotion means promotion of both imports and exports. If a country focuses only on export promotion, it implicitly requires that other countries increase their imports without increasing their exports. It is a special form of "beggar thy neighbor" policy. Hence, such policies are not in the interest of the neighboring countries and are not suited for promoting intraregional trade. Moreover, export-promotion policies that aim at supporting the export sector implicitly tax the other sectors and may distort the domestic economy even more. It should be noted that in some cases, export-promotion policies are pursued to correct for existing distortions due to taxation of the export sector. Such policies will not be discussed in this section because their outcome on trade is unpredictable. Instead, the focus is on the trade-promotion policies that free the internal economy from distortions and aim to increase trade in commodities.

As noted earlier, the exchange-rate problem is at the heart of the trade problem. Hence, agricultural trade performance depends very much on macroeconomic stabilization policies that may eventually result in the abolition of the foreign-exchange allocation system and the introduction of convertible currencies. However, it may take a long time to achieve such a situation; meanwhile, the agricultural trade environment can be improved.

JOINT ACTIONS BY A GROUP OF COUNTRIES

It should be quite clear that the individual countries should have a strong self-interest in reforming their policy, regardless of what the other countries do. Nevertheless, an internal trade policy reform may become more effective if similar actions are carried out by neighboring countries at the same time. Past experience with regional integration schemes indicates that individual countries unrealistically expect to solve their foreign-exchange problem through regional cooperation, and hence they postpone internal policy reforms. A more realistic approach would be for subgroups of two or more countries to cooperate whenever they perceive mutual benefits. "Such an incremental approach should not involve further proliferation of organizations but should involve bilateral (or multilateral) agreements between governments that perceive benefits from a mutual liberalization of product and factor markets" (World Bank 1989). Actually, it might be a reasonable approach to start the reform process with this form of regional cooperation.

Harmonization of Price Movements and Price Levels

There is ample evidence that an isolated national food price policy is very costly. When prices in one country are significantly lower than in neighboring countries, they provide an incentive for illegal transborder trade. Such trade flows may even lead to a breakdown of the internal stabilization and liberalization policy. Take, for example, the case of Zambia. It is well known that low prices for maize meal in 1989 stimulated smuggling to Zaire. Low maize prices in 1990/91 led to shipments of white maize to Malawi and Zaire on a large scale and of yellow maize to Zimbabwe. Furthermore, depressed domestic prices certainly contributed to the lower harvest in 1991. It should not be surprising that a maize shortage in Zambia materialized in late 1991. Illegal fertilizer exports from Zambia to the neighboring countries, mainly to Zaire and Malawi, had the same effect. According to estimates, the annual cost to Zambia of such cross-border leakages of fertilizer in recent years probably range around 75-150 million kwacha (Kinsey 1991). This supports the hypothesis that divergent national food and agricultural policies can be counterproductive in achieving food security on the national level, and, moreover, countries lose in welfare. It is often argued that unofficial trade is generally welfare-generating, and governments may be well advised not to suppress such trade (Roemer and Jones 1991, 217). This may indeed be true in some cases, but between Malawi, Zambia, and Zimbabwe, illegal trade often implies trade deflection. The country with lower consumer protection or even subsidization can import from countries outside the region and export to neighboring countries. Thus, the country not only incurs a financial loss but also a loss in overall welfare. Even in other more favorable cases it is obvious that substitution of legal trade for illegal trade increases welfare, mainly because transaction costs are lower.

Tables 3.10, 3.11, and 3.12 in Chapter 3 illustrate the differences in producer and consumer prices between the countries.

Adjusting prices would not only remove incentives for illegal transborder trade but would provide the basis for trade expansion.

Coordinated Removal of the Foreign-Exchange Constraint

Certainly, the solution of the foreign-exchange constraint is primarily a concern of internal policies. Nevertheless, internal policy adjustments would be more efficient if a group of countries cooperated.

One strategy could be to strengthen the role of the existing Preferential Trade Agreement (PTA) between the eastern and southern African countries in trade and exchange-rate activities; another could be for individual countries to coordinate their national policies on trade and exchange rates.

The PTA introduced a scheme that requires clearing of trade imbalances among the partner countries in hard currencies. Success so far has been limited. The scheme is built on the hypothesis that balanced trade among the PTA members or any smaller group of countries is beneficial to the trading partners. However, that is not necessarily true if the currencies of the trading partners are overvalued to a different extent and if external tariffs differ among the trading partners: countries with weaker currencies will benefit at the expense of countries with stronger currencies (Koester 1986, 63).

However, there might be an alternative. Countries might be inclined to set up a list of products for which private cross-border trade is allowed. A commodity should be put on the list only if the corresponding domestic market has been decontrolled. Such a condition has been met, for example, for most products in Zambia from June 1989 onward. A second condition for mutual beneficial exchange of commodities is a relatively small margin in the extent of the overvaluation of the currencies among the three countries studied. If these conditions were fulfilled, countries could accept trading partners' currencies, assuming that the price received in these markets is higher than the price the country would get in hard currencies in the international market. This approach is based on the perception that trade between neighboring African markets is more beneficial than trade between African and overseas countries because transport costs are lower within the region. The gain from lower transport costs may be higher than the loss incurred from accepting overvalued currencies. Nevertheless, countries could start a list of products to be traded if the overvaluation of their corresponding currencies did not differ greatly from each other and if the gap between import and export parity prices for the listed products was fairly large. Quiroz and Valdés (see Chapter 4) found that the exchange-rate distortion of Malawi, Zambia, and Zimbabwe differed only by 4-8 percent from 1980 to 1987 and has most likely declined thereafter. The first condition being met there is certainly a large set of products that could be placed on the list after internal market liberalization.

Coordinated Management of Food Aid

Food aid can have negative effects on the willingness of countries to engage in commercial trade. If food aid, which may be urgently needed by some countries, leads to an inflow of food from outside the region, it disturbs the integration of markets. However, food aid could be used to promote trade, and moreover, trade could be used to make food aid more efficient for the region as a whole. These results could materialize if food aid were to be given only in the form of "triangular" purchases involving a donor and either two aid recipients or a commercial importer and one food aid recipient. Actually, triangular purchases have increased in recent years, and most grain trade in the region is based on this type of transaction, but the past record indicates that triangular transactions could be improved.

First, the type of food aid received must match the type of grain in deficit in the region. Hence, donors should provide food aid on the basis of regional shortfall and not on the basis of a specific country's production shortfall in one specific type of grain.

Second, the timing of this transaction within a given year could be improved. The Food Security Unit of SADCC has set up an efficient early warning system that provides information on predicted production early in the year. The countries could also make use of their grain marketing boards more efficiently. Most of the grain is bought-in within a short period of the harvest (Koester 1990, 66). Hence it is possible to determine quite early in the year what the actual harvest will be. Therefore, governments are also able to contract purchases early in the year, preferably before the harvest in the neighboring countries has been bought-in. Donors could be approached early in the year, and food aid should be brought in only to close the gap of production shortfall on the regional level. External assistance can be provided, of course, in the case of production shortfalls in individual countries, but it does not have to take the form of food aid.

Coordinated management of food aid would also help to ease one of the major obstacles to intraregional trade, the transport constraint.

Bilateral or Trilateral Trade Agreements

Bilateral trade agreements are occasionally set up even in trade relations among industrialized countries. In most cases, they are either the consequence of protectionism, which may lead to trade preferences, or of state-controlled trade like the agreements between the United States and the former Soviet Union.

Trade between Malawi, Zambia, and Zimbabwe was, until recently, completely conducted by parastatals, and it might be worthwhile to investigate the scope for bilateral or trilateral agreements.

Agricultural markets in the three countries are very much dominated by the governments' objective to maintain food security. Governments do not want to rely on the vagaries of external markets. This attitude leads quite often to a situation where an excellent harvest is nearly as much a national disaster as a serious production shortfall. In the first case, governments may have to stock up huge amounts of food and may be in need of increased storage capacity and finances; in the second case, they may be short of foreign exchange to pay the additional import bill. The situation of the individual countries could certainly be improved if countries were to coordinate their stockholding (see Chapter 8) or if they could rely on deliveries from neighboring countries. Thus, bilateral or trilateral agreements could include the commitment of one country to deliver a specific quantity of a specific commodity and the commitment of another country to buy that quantity. The agreement should—as in most other trade agreements—include only the specification of the quantities to be traded. Prices to be paid should reflect the prevailing international prices at the time of the transaction.

A wide range of products could be included in such trade agreements. The first group of products could include the commodities of which one country is or could be a net exporter and neighboring countries are net importers. Thus, this agreement would merely redirect trade flows. Actually, there is a wide range of products that could be included in such a list (see Table 13.1), even with the present pattern of production and consumption. It should be noted that for some product categories, such as agricultural tools, there would be a scope for intraindustry trade.

Of course, it is not at all advisable to include all the products in a bilateral or trilateral agreement. Some of these products might instead belong to the group of products for which free trade is allowed between the three countries.

Table 13.1—Main agricultural products to be traded between Malawi, Zambia, and Zimbabwe

Product	Exporter	Importer
Yellow maize	Zimbabwe/Zambia	Malawi
Poultry	Zimbabwe	Zambia/Malawi
Dairy products	Zimbabwe	Malawi/Zambia
Fish/fish meal	Malawi	Zimbabwe
Vegetables	Malawi	Zimbabwe
Cotton	Zimbabwe/Zambia	Malawi
Seeds	Zimbabwe	Malawi/Zambia
	Malawi	Zambia/Zimbabwe
	Zambia	Malawi/Zimbabwe
Fruit seedlings	Zimbabwe	Malawi/Zambia
Cattle genetic material	Zimbabwe	Malawi/Zambia
Fertilizer	Zimbabwe	Malawi/Zambia
Agrochemicals	Zimbabwe	Malawi/Zambia
Agricultural tools/ implements	Zimbabwe Zambia	Malawi/Zambia Malawi/Zimbabwe
Agricultural equipment/ machinery	Zimbabwe	Malawi/Zambia
Tractor spares	Zimbabwe	Malawi/Zambia

Source: Based on data from J. Rusike, *Traders Perceptions on Constraints on Expanding Agricultural Input Trade Among Selected SADCC Countries*, Working Paper AEE 5/89 (Harare: University of Zimbabwe, Department of Agricultural Economics, 1989); Chr. Michelsen Institute, SADCC Intraregional Trade Study, Main Report (prepared for SADCC by the Department of Social Science and Development, Bergen, Norway, 1986); U. Koester, "Agricultural Trade Among Malawi, Tanzania, Zambia, and Zimbabwe" (paper prepared for the World Bank, Agricultural Division, Southern Africa Deparment by the International Food Policy Research Institute, Washington, D.C., 1990); Food and Agriculture Organization of the United Nations, *FAO Trade Yearbook* (Rome: FAO, 1990).

Trade in white maize poses specific problems. Zimbabwe is certainly able to produce an exportable surplus in normal years. However, fluctuations in production are more pronounced for this country than for the other two. Hence, if Zimbabwe wants to become a reliable supplier, even in years of drought, it either needs huge stockpiles or has to be prepared to import from countries outside the region to meet its commitments in a drought year. Such a strategy may not look very profitable at first glance, but it could be viable. The risk involved could be reduced if the internal market were allowed to adjust to changes in production and world market prices.

Joint Development of a Regional Transport, Marketing Infrastructure, and Communication Network

Exchange of goods among economic agents depends very much on transaction and transport costs. Transaction costs include all cost items that are related to the exchange of commodities, except transport costs.

The main elements of transaction costs are (1) costs incurred in searching for trading partners; these costs will be high if there are restrictions on travel, if the communication network performs poorly, if markets are not organized, and if products are not standardized, and (2) costs incurred to enforce contracts; these costs also depend on restrictions on travel, but additionally on the functioning of the legal system.

In a free market system, transaction costs and transport costs account for the price differential between the price buyers have to pay and the price sellers receive. Hence, the higher these costs, the more it pays for an individual not to become involved in the exchange of commodities but to remain a subsistence producer. The same holds true for trade between countries. Thus, trade promotion should aim at reducing transport and transaction costs. It is plausible that some elements of transport and transaction costs can be lowered if neighboring countries cooperate. First, restrictions on traveling could be eased. Second, the transport and communication network could be enlarged or improved through joint projects among the three countries.

UNILATERAL ACTIONS TO PROMOTE INTRAREGIONAL TRADE

It is in the interest of individual countries to promote trade independently of what other countries do. As mentioned earlier, the best fuel for trade promotion is a policy of external and internal liberalization. However, this can succeed only if the government succeeds in stabilizing the economy and in removing the foreign-exchange constraint. It is true that collective actions could also contribute to the solution of the foreign-exchange problem. However, countries cannot expect to solve their foreign-exchange problem through regional cooperation. The root of the exchange-rate problem lies in internal policies; therefore, these policies have to be changed. Trade promotion is a positive by-product of internal stabilization and liberalization policies. Of course, these policies have to encompass the agricultural sector too.

Synchronization of External and Internal Liberalization Policies

Growth in agricultural trade will follow from external and internal liberalization measures. The two sets of measures have to be synchronized. External liberalization can actually be counterproductive if not supported by internal measures. External liberalization may not contribute to an increase in overall welfare if the internal economy is not allowed to adjust in the right direction. To achieve this, some internal liberalization is needed simultaneously. As for stabilization and external liberalization policies, it would be advisable for internal liberalization policies to be gradually phased in. A few of the necessary steps will be discussed in the next sections.

Changes in Price Policies

Individual countries should have a genuine self-interest in price adjustment. Actually, this could be done without too many changes in other areas. Most prices in Malawi and Zimbabwe and maize and fertilizer prices in Zambia are set by the governments. The price-setting criteria are based on internal considerations, such as achievement of food security and cost of production. Hence, it is not surprising to observe a wide variance in price levels among the countries and even a variance in the movement of prices over time. If policymakers are advised to give up the past price-setting mechanism, they should be convinced that, on the one hand, food security and the welfare of the society at large will be improved and, on the other hand, that they will not lose political support. As the maize economy is still the most important sector from the political point of view, the effect of recommended maize prices will be discussed in more detail.

Table 3.11 in Chapter 3 shows that relative prices differ significantly among countries. Zimbabwe provides higher incentives to producers of coffee, tobacco, and tea than the other countries do, even though these exportables are taxed in all countries because of their overvalued currencies (Koester 1990, 46). If the other countries were to accept the price pattern of Zimbabwe, one could expect a decline in their maize production, lowering the domestic level of self-sufficiency.

However, the countries would at the same time be able to earn more foreign exchange for their main export crops, allowing them to import the needed quantities of maize. It might well be that policymakers in the region have not followed this path so far because they are afraid that the shortfall in foreign exchange would not allow them to import the needed maize, or they may fear that because the white maize market is a thin market in the region, insufficient amounts of white maize would be available. These concerns may have been justified in the past, but will be less so in the future.

First, if the countries cooperate in the way discussed above, the three countries together will most likely produce more, not less, maize, and the trading agreements should give individual countries access to as much maize as needed. This emphasizes the importance of synchronizing unilateral and collective actions on the one hand and internal and external policy changes on the other.

Second, the emergence of the yellow maize economy has built some stabilizing factors into the overall maize economy. This is partly due to the higher production stability of yellow maize but also, according to a new study, to the fact that yellow maize may be blended with white maize to a certain extent with actually positive nutritional effects and no difference in taste.[1]

Maize production in the individual countries would also be affected if relative prices of other grains were to be adjusted.

It is not possible to accurately quantify the trade effects of a realignment in prices because own-price and cross-price elasticities are not available. However, it is known that farmers react to changes in price incentives. Moreover, even small changes in production may give rise to significant changes in trade flows among these countries because all of them are nearly self-sufficient in the main grains, apart from wheat and rice.

Policymakers should note that the effects discussed could also follow from an adjustment in relative prices. For example, the price of maize could even be used as a numeraire, imposing changes only on all other prices. This would not, of course, allow the countries to capture the positive trade effects arising from adjusted maize prices across the countries. However, it would have the positive effect of reducing the fear that price adjustments impair food security. Hence there is a strong

---

[1] According to the author's verbal communication with Jim McKenzie, Ministry of Agriculture, Zambia's consumers do not notice if yellow and white maize are mixed, up to 30 percent of yellow maize.

argument for realigning relative prices.

Governments should be more concerned about changing the level of agricultural prices. As shown above, the agricultural sector is taxed under the present policies. Hence, the agricultural price level should go up in all three countries if prices were liberalized. The agricultural sector would gain as a whole, but for the urban population the outcome may be negative if changes in producer prices are transmitted to consumer prices. However, one should have in mind that the price adjustment does not merely redistribute income from the urban to the rural population. Instead, the main consequences will be an improvement in the efficiency of the economy, an increase in the availability of foreign exchange, and thus an increase in the availability of consumer goods and inputs.

Changes in setting agricultural prices should be instituted as soon as possible, but price reform has to include much more. Governments should eventually give up setting prices at all and should rely only on some stabilization measures that will be discussed below. As soon as possible, regional and seasonal price variances should be allowed to reflect market forces. It is not necessary at this point to expand on the rationale for such a policy step, because much has been written on the subject. However, the ongoing policy debate should take into account some new research findings.

First, in most cases it is difficult to enforce uniform regional and seasonal prices, even in countries like Zimbabwe with a functioning Grain Marketing Board (Chisvo et al. 1991, 128). Furthermore, prices in the parallel market fluctuate, and the more regulated the official market is, the greater the fluctuation. Research from other countries suggests that it is the small farmer and the poor consumer who are confronted with the instability of the parallel markets. Hence the highly regulated markets actually may contribute indirectly to the food insecurity of the poor households.

Second, there is new evidence that panseasonal and panregional prices increase food insecurity for rural grain-deficit households (Jayne, Chigume, and Hedden-Dunkhorst 1990).

Third, the transport system in the region is under increasing strain. There is strong evidence that the present market regulations put significant pressure on the transport system for the following reasons.

- Production has been stimulated in regions far from the consumer markets (Jansen 1988).
- As the present price policies imply a higher subsidy for crops in distant locations, market regulation favors production of high-volume crops in these regions.

- Nuppenau's research (see Chapter 12) indicates that savings on transport costs could amount to more than 12 percent if a more efficient routing of trade flows were permitted.
- The present system of panseasonal prices requires that the grain be sold to the public mills soon after harvest, since there is no price incentive for the farmer to hold the grain. Later in the year, transport is needed to move the meal to the rural areas. This transport pattern means that more often than not, trucks are empty on the return trip.
- The use of more traditional modes of transport would relieve some of the strain on the modern transport system. Studies on parallel markets indicate that nonmodern transport modes are often used.
- By easing the strain on the transport system in the agricultural sector, policy reforms of the present agricultural market organization would free the constraints on other sectors of the economy.

Some Pitfalls in Implementing Price Reform

It is understandable that governments are reluctant to implement price reform in one stroke. Indeed, the three countries would be well advised not to do so. The phasing-in of the price reform allows countries to develop the institutional framework needed for the functioning of a market system and reduces the risk of the reform, because the performance of the economy can be closely monitored. All three countries have already had some experience in reforming their price policy, consequently some lessons can be drawn.

First, politically set prices should be adjusted to reflect export or import parity prices. This can be done without changing the policy framework to a large extent.

Second, governments have tended to liberalize prices step-by-step for specific sets of commodities. Prices of maize or maize meal and fertilizer (Zambia may serve as an example) rank low on the list of liberalized product markets. Such a strategy can easily lead to highly distorted markets and even to higher food insecurity. Both Zambia and Zimbabwe had mixed experience with partial liberalization. Zambia liberalized all product prices except white maize, maize meal, and fertilizer in 1990. Depressed maize prices led to a flow of illegal maize exports across the border. It turns out that Zambia may have to import about the same quantity commercially that had been exported illegally—obviously a very costly experience for the country. Zimbabwe had a similar experience with the liberalization of the yellow maize market. The objective was to give an incentive to allocate more acreage to white maize, but the outcome was the opposite. This clearly demonstrates the need to recognize the interdependency of markets.

Third, changes in domestic price policies have to be tuned to changes in external trade policies. Zambia's experience this year with allowing free transborder trade in yellow maize can serve as an example. As yellow maize was allowed to cross borders legally, and prices for this type of maize were higher in Zimbabwe, the availability of yellow maize in Zambia declined. However, as it was forbidden to feed white maize to animals, market prices for yellow maize increased and surpassed prices for white maize. Clearly, this exerted an incentive to feed white maize to animals—even if not legally permitted—and, thus contributed to the white maize shortage that the country will face later this year.

Fourth, liberalizing prices may be counterproductive if the marketing system is not prepared to provide marketing services efficiently. The experience of Zambia illustrates this point. The harvest of the marketing year 1990/91 was negatively affected because by the end of December only 69 percent of total requirements for basal dressing fertilizer had been supplied, while the supply of top dressing fertilizer stood at 22 percent. Maize seed supply was only 80 percent of requirements (SADCC 1990).

Changes in the Marketing System

It has been generally accepted that the governments (including parastatals) should play a different role in monitoring the marketing of agricultural products than they have in the past.

Privatization of the marketing sector should rank high on the agenda. However, privatization should not be narrowly considered as a goal in itself. The ultimate objective should be to improve the economic efficiency of the marketing system and to contribute efficiently to some noneconomic government objectives. In general, it can be expected that private agents in a competitive environment may increase the efficiency of the marketing sector, because they have to adjust quickly to changes in market conditions if they want to survive. However, complete and abrupt privatization may result in worse market performance. On the one hand, it may not provide a competitive environment, because only a few private traders may have the necessary skill and capital (including transport means) to engage in marketing. On the other hand, a private marketing system demands not only private traders but also an adequate marketing infrastructure, such as a marketing-information system, developed regional markets, private storage facilities, and institutions to monitor the performance of the markets and the quality of products. The development of marketing skills by private traders and of a marketing framework requires time. Timing and sequencing of policy reforms are therefore crucial.

Movement Restrictions

Maize marketing is highly regulated in the three countries (most highly in Zimbabwe). Most of the maize has to be delivered to and bought from the Grain Marketing Board (GMB). The only exceptions concern trading of yellow maize within a specified area around the selling farm and within communal areas. Hence, there is already limited competition for the GMB. Private traders are already active; their functions need only to be gradually extended. Of course, the price policy must be changed for private trading to be profitable. Subsidization of consumer prices and the GMB depresses market prices and provides little incentive for private traders. Nevertheless, the government of Zimbabwe was well advised to lift restrictions on the movement of white maize in the communal areas for the 1992 marketing year (Mangwende 1991). The movement of maize between noncontiguous communal areas will also be allowed. In addition, the government plans to eliminate all existing restrictions on crop movement. This announcement may please those who favor liberalization. However, the experience with the partial liberalization of the trade in yellow maize in Zimbabwe indicates that a change in pricing policy and a change in the role of the GMB should be incorporated simultaneously.

The Future Role of the Marketing Boards

The marketing boards in the three countries were supposed to implement official price policies for key agricultural commodities and to contribute to the development objectives of the governments. As price policies changed (in Malawi since 1986 and in Zambia since 1990), the economic environment for the marketing boards has changed; now they at least have to compete with some private traders. Zimbabwe, which has not yet started this process, can draw on the experience of the other countries. As one of the papers presented at this conference deals with the GMB in Zimbabwe, the focus here will be on the role of the boards for individual commodities during the adjustment period as well as during the final stage. Some suggestions on the change in marketing of white maize will also be offered.

The marketing system of export crops could be privatized the fastest. However, privatization may not improve the efficiency of the marketing system, as the board, even if privatized, may retain monopolistic market power. Barriers to entry may be too high for potential competitors: exporting requires specific skills and facilities and may be profitable only when the quantity traded is large. Hence, alternative measures should be introduced. One possibility could be to open the border to the other two countries. Allowing for trade in export crops among these countries would stimulate competition between the marketing agents in the individual countries.

Another group of commodities, which are not traded on the international market and are not crucial for food security, should also be liberalized in the initial phase of the process. However, to help stabilize these markets, it may be reasonable to first develop the markets by allowing private trading and to privatize the board later.

Changing the marketing system for white maize is the most difficult task. The government and the marketing board need to play a specific role in the maize market.

First, by functioning as a buyer and seller of last resort, the marketing board could buffer the potential price instability that would result from liberalizing the maize market.

Second, the government may want to institute special welfare programs for those who are hurt the most by the new policies, or who can adjust the least. The government may also be inclined to protect from the adverse effects of the policy reforms those groups that are well organized and politically strong. The GMB could be the appropriate institution to implement these programs.

Third, the government may be willing to hold a certain amount of maize as a strategic reserve, and the GMB may serve as a storage facility.

Finally, if a country wants to become a permanent exporter but is not prepared to liberalize external trade in maize, the GMB could function as the main or even the only exporting agency. As noted previously, in order for the country to become a reliable exporter, it must hold large enough volumes of stocks for trade to be profitable in the short run. The GMB may have to hold these stocks. The GMB cannot fulfill these nonmarket objectives without governmental support. Hence, there are only two alternatives—either the GMB remains a public enterprise and the government covers the costs for the non-market-related activities, or the GMB will be privatized and the government will have to negotiate with the GMB the price it has to pay for the GMB's contribution to the governmental objectives. It might not be a reasonable strategy to move from the first alternative to the second in the first phase of the policy changes. Nevertheless, even then the performance of the GMB would have to be monitored continuously to ensure that it performs its marketing functions as efficiently as the private sector.

Independently of the future role of the GMB in the maize trading economy, less strongly regulated maize prices and private trade in maize could be allowed in the first stage of the liberalization process. Governments might be afraid that markets may become more volatile if maize prices are determined by pure market forces, and the GMB may have to play a role in decreasing the market instability. However, this concern will most likely not materialize, at least not in the medium and long terms.

The experience of some Eastern European countries may be of interest. Some of these countries privatized their marketing system in one stroke even when no private trading had been allowed before. The evidence of Poland and East Germany indicates that in the first months after the change, regional markets were not integrated and prices in one district were not linked to prices in other districts. However, markets became more and more integrated from month to month. Private traders very soon collected information on market conditions in districts other than those in which they originally operated and took advantage of price spreads, thus contributing to movements toward uniform prices. It has to be noted that this outcome materialized in an environment where the communication network works much worse than in most developing countries and where the experience of private traders was extremely limited.

Nevertheless, if governments are not prepared to accept the risk of more volatile markets, especially during the transitional period, they may use the GMB to stabilize the markets. However, it should be accepted that uniform prices across time and space conflict seriously with a market economy. Hence, the governments have to accept from the outset that a certain variance in regional and seasonal prices is needed. Moreover, annual prices should reflect somewhat the internal market conditions and the changes in world market prices. A scheme that would allow the development of a private marketing system and at the same time take care of the governments' concerns could include the following elements.

The governments would announce a target price for white maize that would result in a year of normal harvest and would take into consideration a planned amount of export quantities. However, the actual market price may vary by 10 percent. It would be higher if the outcome of the harvest is lower than expected or if world market prices are higher than expected. It is true that such a scheme would introduce more instability in official prices, but actual prices received by farmers may be more stable than they were in the past when farmers sold to unofficial markets. Moreover, producer prices that vary inversely with the outcome of the harvest would likely contribute to greater stability in farm revenue and thus reduce the risk for farmers. Actual prices would be lower if the outcome of the harvest is larger than expected or if world market prices are lower than anticipated. The price variance would not only contribute to a reduction in farmers' risk, but would also revitalize the private storage economy. The expected profit of private stockholding depends very much on expected prices. If prices are expected to be stable over time, there is certainly no incentive to hold working stocks, seasonal stocks, or carryover stocks. Moreover, the introduced price variance would also contribute to a variance in consumption and thus would imply a built-in stabilizing element.

The task of the GMB could be to serve as a buyer of last resort. The GMB should buy maize at prices within the price band in close cooperation with the government. The buying-in price should be uniform over the country and during the season, implying that the GMB would have to buy mainly at the beginning of the harvesting season and in surplus regions far away from the main consumption areas. Actual market prices would be expected to be above the buying-in price in most regions and in most months of the marketing year. Thus, there could be sufficient incentives for private traders and private stockholders.

Measures to Promote Private Marketing

A marketing system based on private agents cannot work efficiently from the outset without governmental assistance. The government should set up as early as possible a marketing-information system to provide information for traders, producers, and grain processors. Moreover, the experience of Zambia and Malawi has shown that private trading may be severely restricted by lack of capital and limited access to credit (Christiansen and Stackhouse 1989). Hence, the governments should contribute to easing this constraint. In addition, the functioning of a private marketing system could be improved if the establishment of hammer mills and of storage capacities was supported.

SUMMARY

Specific policies to promote trade among a group of countries are needed if the potential for trade exists and can be exploited when specific obstacles are removed.

Agricultural trade promotion should be undertaken only as a part of an overall liberalization and structural adjustment policy. Hence, this study presents a broad approach to policy reform before it focuses on policies directed at agriculture.

Agricultural trade promotion as an integral part of overall policy changes will succeed only if the governments are able to implement sustainable policies. These policies are not necessarily the best ones from a purely economic point of view. Whether the public accepts policy changes depends very much on a government's ability to convince its constituents that a change in policies is unavoidable. In this context, the study presents some rationale for changes in policies, discusses the role of the government, and outlines some guidelines for phasing-in agricultural trade promotion policies.

Even if individual countries have a strong self-interest in reforming their policies, internal trade policy reform is made more effective if carried out simultaneously by neighboring countries.

Joint actions by the three countries in this study should first of all include harmonization of price movements and price levels. There is ample evidence that it is very costly for individual countries to pursue an isolated national food policy.

The countries cannot be expected to remove the foreign-exchange constraint in the very short run. The study presents some alternatives. For example, to the extent that the overvaluation bias of their exchange rates does not differ greatly between countries, many products may be traded freely in the region.

Food aid can be a major obstacle to intraregional trade. However, food aid used in "triangular" trade activities could be used to promote trade among the countries in the region.

The study also explores the scope of bilateral and trilateral agreements to promote trade and suggests groups of commodities to be included in such agreements.

Individual countries would be badly advised to expect regional cooperation to remove the exchange-rate constraint. The root of the exchange-rate problem lies in internal policies; therefore, these policies have to be changed.

In designing liberalization and structural adjustment policies, a proper synchronization of external and internal liberalization policies is crucial. The study identifies the changes in internal price policies that are needed in order to exploit the trade potential.

There is a need to change the marketing system. However, even if privatization of the marketing sector should rank high on the agenda, the author argues that not all the activities of the marketing boards should be privatized in the early phase of the liberalization and structural adjustment policy; for example, the market for maize would not be completely privatized in the foreseeable future.

The GMB should have a specific role in marketing maize, including stabilizing prices, holding a strategic reserve, and performing international transactions in maize. Consequently, if the marketing board has to fulfill some nonmarket objectives, it cannot function as a private enterprise without governmental support.

REFERENCES

Chisvo, M., T. S. Jayne, J. Tefft, M. Weber, and J. Shaffer. 1991. Traders' perception of constraints on informal grain marketing in Zimbabwe: Implications for household food security and needed research. In *Market reforms, research policies and food security in the SADCC region*, ed. M. Rukuni and J. B. Wyckoff. Proceedings of the Sixth Annual Conference of Food Security Research in Southern Africa, November 1990. University of Zimbabwe/ Michigan State University Food Security Research Project, Department of Agricultural Economics and Extension, Harare.

Chr. Michelsen Institute. 1986. *SADCC intraregional trade study*. Bergen, Norway: Chr. Michelsen Institute.

Christiansen, R. E., and L. A. Stackhouse. 1989. The privatization of agricultural trading in Malawi. *World Development* 17 (5): 597-599.

FAO (Food and Agriculture Organization of the United Nations). 1990. *FAO Trade Yearbook*. Rome: FAO.

Hawkins, J. J., Jr. 1991. Understanding the failure of IMF reform: The Zambian case. *World Development* 19 (7): 839-849.

Jansen, D. 1988. *Trade, exchange rate, and agricultural pricing policies in Zambia*. A World Bank Comparative Study. Washington, D.C.: World Bank.

Jayne, T. S., M. Chisvo, B. Hedden-Dunkhorst, and S. Chigume. 1990b. Unravelling Zimbabwe's food insecurity paradox: Implications for grain market reform. In *Integrating food and nutrition policy in Zimbabwe*. Proceedings of the First Annual Consultative Workshop on Food and Nutrition Policy, ed. T. S. Jayne, M. Rukuni, and J. B. Wyckoff. University of Zimbabwe/Michigan State University Food Security Research Project. Harare: Department of Agricultural Economics and Extension.

Kinsey, B. H. 1991. A regional study of demand, supply and distribution and the potential for intraregional trade in fertilizer: Malawi, Zambia and Zimbabwe. International Food Policy Research Institute, Washington, D.C. Mimeo.

Koester, U. 1986. *Regional cooperation to improve food security in southern and eastern African countries*. Research Report 53. Washington, D.C.: International Food Policy Research Institute.

_____. 1990. Agricultural trade among Malawi, Tanzania, Zambia, and Zimbabwe. Paper prepared for the World Bank, Agricultural Division, Southern Africa Department. International Food Policy Research Institute, Washington, D.C., Mimeo.

Lal, D. 1987. The political economy of economic liberalization. *The World Bank Economic Review* 1 (2): 273-299.

Mangwende, W. 1991. Policy statement for the 1991/92 agricultural production year. Speech delivered in August 1991 in Harare, Zimbabwe.

Nelson, J. M. 1984. The political economy of stabilization: Commitment, capacity, and public response. *World Development* 12 (10): 983-1006.

Roemer, G. M., and C. Jones. 1991. What have we learned about policy? In *Markets in Developing Countries: Parallel, Fragmented, and Black*, ed. M. Roemer and C. Jones. San Francisco, Calif., U.S.A.: ICS Press.

Rusike, J. 1989. *Traders perceptions and constraints on expanding agricultural input trade among selected SADCC countries*. Working Paper AEE 5/89. Harare: University of Zimbabwe, Department of Agricultural Economics.

SADCC (Southern African Development Coordination Conference). Regional Early Warning Unit for Food Security. 1990. *SADCC Food Security Bulletin*, October-December. Harare: SADCC.

Sahn, D. E., and A. Sarris. 1991. Structural adjustment and the welfare of rural smallholders: A comparative analysis from sub-Saharan Africa. *The World Bank Economic Review* 5 (2): 259-289.

Stanning, J. L. 1989. Grain retention and consumption behavior among rural Zimbabwe households. In *Household and national food security in southern Africa*, ed. G. Mudimu and R. H. Bernstein, 80-102. Proceedings of the Fourth Annual Conference on Food Security in Southern Africa, 31 October-3 November, 1988, Harare.

World Bank. 1989. *Sub-Saharan Africa: From crisis to sustainable growth*. Washington, D.C.: World Bank.

_____. 1990. *Malawi through poverty reduction*. Washington, D.C.: World Bank.

_____. 1991. *World development report: The challenge of development*. New York: Oxford University Press for the World Bank.

14
Reflections on the Impact of Recent Trade and Political Liberalization in Southern Africa

Kay Muir-Leresche

In this chapter, an attempt is made to take a broad view of the impact of the recent economic and political changes in southern Africa on the future of agriculture generally and agricultural trade in particular.

The situation in South Africa is widely assumed to be on the point of resolution, with a free and peaceful South Africa/Azania anticipated before the end of this century. While there is still considerable doubt that there will be full majority rule before 2000, some of the probable regional impacts on agricultural trade of a fully democratic South Africa are considered here.

The structural adjustment programs being embarked upon within Malawi, Zambia, and Zimbabwe have the potential to affect agricultural development, food security, and trade flows in the region more than the political changes in South Africa. Protectionist policies, including trade policies and the artificially determined foreign-currency, producer, and consumer prices, have restricted regional trade far more than intercountry political barriers. South Africa is the dominant trading partner for most countries in the region despite the political situation, and although resolving political issues will make communication easier, allow greater market integration, and have important implications for regional stability and for skills transfer and institutional development, it is less likely to directly increase or change trade patterns.

Domestic politics has, on the other hand, contributed significantly to the existing distortions, and the removal of these distortions could increase the opportunities for regional trade and development. Insofar as the political imperatives within the countries continue to be seen as appeasement of an urban elite and the desirability of maintaining a situation in which individuals can accrue high rents from the distortions, the decontrol will not actually take place. Only lip service will be paid to the program, with measures taken far enough to attract the necessary international finance before finding it "politically impossible" to continue. There are many dedicated and determined technocrats and politicians who recognize the importance of carrying through the changes, but it is uncertain that they have the political power to ensure that barriers to competition are lifted in time to guarantee the success of these programs.

In this chapter, however, it is assumed that the structural adjustment programs will actually go ahead and will reduce barriers to competition and trade, resulting in market prices that more closely reflect opportunity costs.

If South Africa effectively liberalizes trade and reduces farm subsidies, and the Southern African Development Coordination Conference (SADCC) countries also liberalize trade while at the same time reducing the various indirect agricultural sector taxes, South Africa could become an important market for regional agricultural surpluses. Such policies could see agricultural production expanded. For example, Zambia has a considerable vent for surplus, provided it has the markets and can attract investment into the production of maize. An effectively implemented liberalization program could encourage the movement of capital and skills to these countries, but only if they have succeeded in creating healthy investment environments.

The expansion of these export markets would, in turn, generate the foreign currency to purchase imports, which, if directed toward inputs, could result in increased local industrialization. There is already considerable competition between the SADCC countries to attract capital and skills, and the opening of South Africa would exacerbate this situation. It would also raise questions about the role of international aid tied to products from the donor country when the newly reconstituted region could provide these—for example, competition between donor suppliers of agricultural equipment and South African suppliers.

Historical and Theoretical Background

Most of the countries in the region have in common their colonial background and their emergence to independence in the 1960s. They were characterized by poorly developed and highly skewed infrastructure and institutions, with the Portuguese-dominated countries having particularly poor records in social, infrastructural, and economic development but better racial integration. Three countries (South Africa, Namibia, and Zimbabwe) in the region have had significant settler communities, which resulted in far more effective development of the resources in those countries than in the direct colonies, but this development was highly skewed toward the white populations and resulted in racial segregation and white dominance politically and economically. This has placed a far greater burden on the emergence of independent, majority-ruled governments in these three countries, involving political revolution and international economic sanctions. Zimbabwe and Namibia are now independent, and South Africa is currently moving toward greater racial integration, but interpretations of majority rule and democracy differ widely between the main protagonists.

Most of the countries in southern Africa have displayed a tendency toward totalitarian governments that operate highly regulated, noncompetitive economies. Some countries have allowed political opposition, but they have generally been ineffective because the inherited infrastructure and institutions have lent themselves to the exchange of one elite for another. Recent changes in Eastern Europe and the general move throughout the world toward liberalism in politics and laissez-faire economics has led to some questioning of the status quo throughout the region. This is expected to have positive effects on economic growth and equity.

It is uncertain whether this liberalism follows the utilitarian school that promotes liberal government not for the sake of liberty but simply because of its effects on efficiency. There appears to be some belief that movement toward market-led economies may be desirable but that this liberalism would not necessarily be applied to the political arena. However, in order for a liberal government to operate efficiently there must be a liberal society (Mill 1859). There also remains the eternal conflict between economic efficiency that may reinforce or extend existing inequalities and growth through redistribution. In fact, in many of these countries the markets are so distorted that the removal of many regulations may actually increase both equity and efficiency. To what extent the new liberalism should, will, or can combine the capitalist system of production with a more socialist system of distribution is still unclear. Much of the policy analysis carried out using welfare economics is an attempt, if not to reconcile these conflicts, then at least to highlight the trade-offs involved so that policymakers are able to make informed choices.

It is unclear whether the new liberalism sweeping the world incorporates the notion that social welfare is a matter of concern to all men or whether it is more egocentric. If the new liberalism looks toward the market to maximize social welfare and does include the notion that individuals have a wider responsibility to humanity, then it will have to find effective ways for the gains from the market to reach a broader spectrum of the population. It is now much more widely accepted that the pursuit of individual welfare maximizes efficiency or social welfare at the existing income distribution levels, provided always that there are no market imperfections. However, at the macro level, during both depressions and inflationary spirals, the good of the individual and the good of society are often in direct conflict—with consumers saving when the economy would be better if they were spending, and vice versa. Given the high inflation levels accompanied by high unemployment common in southern African countries, individual behavior often exacerbates the position. This, together with the various market imperfections, has often been used to justify the widespread and often conflicting government intervention in macro-variables. In most of southern Africa, however, these regulations (which include significant nontariff barriers to trade) have become self-defeating, and there is a strong argument for an abrupt approach to structural adjustment such as was undertaken by Germany in 1948. The main problem with the gradual approach is that the economy remains in disequilibrium.

As the earlier chapters have shown, most of the economies of southern Africa have been very heavily regulated. Government spending and budget deficits are out of control, causing spiraling inflation made worse by the severe restrictions on supply expansion. Many but not all of the regulations restricting supply arise from the shortage of foreign currency. The structural adjustment programs in the region are an attempt to allow market forces the opportunity to reverse the stagnating or negative growth rates. Most of the countries are implementing social programs to reduce the short-term negative effects on the poorest sectors in the community, but effective programs are difficult to achieve. The liberalization of both internal and external markets is an essential component of the strategy, and until barriers to entry are broken down, the economies are unlikely to respond and further entrenchment of monopolies is likely to occur. Freer trade within the region would benefit all the countries within SADCC and would contribute to increased food security in the region.

Effects of a Free South Africa on Agricultural Production and Trade Patterns

It is unlikely that the emergence of a free and peaceful South Africa will of itself make a major difference to agricultural trade within the region. This premise is based upon the existing regional trade patterns, which are already heavily biased toward trade with South

Africa. Within the SADCC region, only Angola and Tanzania have no trade with South Africa. The economies of most SADCC countries are highly dependent on South Africa. This is particularly true of Lesotho, Swaziland, and Namibia, all of which belong to the Rand Monetary Area and which together with Botswana are involved in the South African Customs Union (SACU). The countries are heavily dependent on SACU for government revenue; for example, some 70 percent of government revenue for Lesotho, 60 percent for Swaziland, and 40 percent for Botswana comes from SACU. Migrant labor to South Africa is Lesotho's dominant export, but otherwise South Africa accounts for a lesser percentage of national exports than imports; for instance, some 20 percent of Swaziland's exports but almost 90 percent of its imports. South Africa provides over 80 percent of the import requirements of SACU members—around 40 percent for Malawi and 15-20 percent for Zimbabwe, Zambia, and Mozambique. All of these countries rely heavily on South Africa for transport, ports, electricity, and oil.

A majority government in power in Pretoria will ease psychological and moral barriers but is not likely to increase trade that is already at such high levels. There is a real danger that capital and labor will move out of these countries back to South Africa as trade sanctions are lifted and manufacturers (for example, the Taiwanese textile manufacturers) who located in the BLS countries (Botswana, Lesotho, and Swaziland) to avoid sanctions relocate to South Africa. In addition, there will be less moral pressure on international aid to provide assistance to the countries of southern Africa.

Malawi, Zambia, and Zimbabwe also rely heavily on South Africa as their dominant trading partner. South Africa is Zimbabwe's single most important trading partner. The position on exports fluctuates, with more of Zimbabwe's exports going to the United Kingdom in some years (for example, 1987), but South Africa accounts for almost 20 percent of all Zimbabwe's imports. Although the U.K. is Malawi's principal export market, South Africa is the principal supplier, accounting for some 30-40 percent of imports. It is unlikely that the change in government in South Africa would increase trade links directly. Any expansion in regional trade requires decontrol of domestic and external markets.

The political changes may still have considerable indirect effects on agricultural development in the region even if there is no liberalization. A majority-ruled, SADCC-recognized government in South Africa would have an immediate impact on communication and skills transfers. Agricultural research is an area that would benefit considerably from communication. For example, Zimbabwe would benefit significantly from greater exposure to the agronomic practices for grain production in South Africa, which are geared to lower rainfall and would therefore be suitable to the more arid Natural Regions 3, 4, and 5. Another example is sunflowers, which are an important commodity in South Africa but have been virtually ignored in Zimbabwe. Better access to research and inputs for sunflower production could be particularly important for smallholders in the semi-arid regions of Zimbabwe.

Equally, South Africa has had little experience with development economics, farming systems research, and project development, assessment, and management in rural areas; for example, experience with smallholder credit schemes, cooperatives, market access, and rural development in general. It could benefit greatly from the research and experience of the SADCC countries.

Institutionally, South Africa could play an important role in a number of areas, including the regional food-security framework, particularly through participation in the early warning system. Its potential role as a market for regional surpluses in good rainfall years is almost as important as its potential role in spreading the drought risk. South African rainfall patterns usually differ from those in Zimbabwe, Zambia, and Malawi,[1] which tend to be more heavily affected by the Intertropical Convergence Zone and therefore to follow more similar drought patterns. It also has a large market for grain stockfeeds, which can be useful in reducing the costs of operating foodgrain reserves. As the earlier chapters have shown, the large fluctuations in production, the almost totally inelastic demand, and the lack of supplies of white maize from elsewhere make foodgrain reserves extremely expensive and difficult to operate.

Unless the individual countries (including South Africa) reduce some of the barriers to trade, however, the full potential of the gains from a politically stable South Africa cannot be achieved.

Structural Adjustment—Its Potential Role in Regional Agricultural Development and Trade

Readers are referred to the other chapters for details of the proposed structural adjustment programs and for the potential gains from trade. This chapter simply introduces South Africa into consider-

---

[1] This was not the case in 1991/92 when the worst drought of the century affected all the countries in southern and eastern Africa. Parts of South Africa were less affected but these were not the grain-producing regions.

ation. It is an attempt to highlight some of the issues that may arise.

As the earlier chapters have shown, agriculture has been heavily taxed in most of the countries in the SADCC region.[2] Assuming, therefore, that structural adjustment programs go ahead, they will involve improving the terms of trade for agriculture and, in particular, improving them for labor-intensive export and import-substituting commodities.

South Africa is facing relatively similar problems to those of many countries in the region except that, since its currency has been less overvalued, it may have actually subsidized producers of wheat, maize, stockfeed, and grazing. Direct producer subsidies and rebates peaked in the 1987 harvest year at R 583 million, with maize subsidies at 19 percent of the value of output. The table below shows the direct subsidies and rebates paid to South African farmers and agricultural marketing boards in the harvest years 1980-89—maize and wheat account for over 80 percent of the subsidy in most years.

Harvest Year	Million Rand
1980	256
1981	297
1982	303
1983	478
1984	503
1985	507
1986	345
1987	583
1988	259
1989	198

Implicit subsidies/taxes from protection are not included. No opportunity-cost pricing analyses appear to be available. It would be interesting to carry out border price comparisons.

Despite the subsidies granted to wheat farmers,[3] it is possible that South Africa may have a comparative advantage in wheat production. Tractors, combine harvesters, and other mechanical equipment are freely available to South African farmers and cost almost five times less

[2] There is further evidence of this taxation in Muir-Leresche and Jansen 1991, Jansen 1988, and Muir-Leresche 1984.

[3] Wheat subsidies were still 12 percent of the value of marketed output in 1990, even after the commitment to reduce subsidies.

than they do in Zimbabwe.[4] South Africa is also able to produce nonirrigated summer wheat, as it does not have the same rust and lodging problems experienced in Zambia and Zimbabwe. This makes the wheat cheaper to produce and probably more economic despite lower yields than in Zambia and Zimbabwe. South African wheat does not need to compete for scarce water resources for irrigation, nor does it require the capital investment in irrigation facilities.

It is possible that the region would benefit from South Africa's moving away from maize and into wheat if there is greater investment in maize production in Zambia and in the surplus areas in Tanzania. A preliminary analysis to assist in determining regional comparative advantage in the major commodities under different trade and exchange-rate regimes in the region would be interesting.

Normalizing access to foreign exchange would reduce the price of capital inputs in most countries in the region, but it would be unlikely to reduce them to the South African price levels because of the larger market in South Africa and the lower transport costs involved. It is likely that South Africa will continue to have comparative advantage in capital- and skills-intensive commodities. Zambia, Tanzania, and, in due course, Angola and Mozambique will have the greatest comparative advantage in land- and rainfall-intensive commodities, while Zimbabwe, Malawi, and Swaziland are likely to have the greatest comparative advantage in labor-intensive commodities. However, natural resources, local demand, and management capacity will also affect these, and it is difficult to determine comparative advantage a priori.

Botswana, Namibia, and Zimbabwe would appear to have an advantage in beef and other livestock exports to South Africa. There appears to be a large vent for surplus if South Africa removes nontariff restrictions and does not place high tariffs on livestock imports from the region. Consumer prices for meat in South Africa are considerably higher than they are in most of the region, with prime beef cuts selling at US$6.50 per kilogram compared with US$2.50 per kilogram in Zimbabwe in October 1991. Other areas where realigned prices and deregulation may stimulate exports from SADCC countries to South Africa include dry beans, cotton, oilseeds, and stockfeeds. This

---

[4] For example, at the end of 1990, tractors cost around R 1,800 per kilowatt, and of the almost 200 makes available to farmers, the prices range between R 24,000 for a Kubota B6200 and R 246,000 for a John Deere 4755 MFWD Cab/AC. Combine harvesters cost between R 23,000 and R 115,000 and Mazda 1600 5-gang Rustler pickups are R 24,000.

assumes that South Africa will eliminate nontariff barriers to trade and avoid replacing them with excessive tariffs.

The opening up of domestic markets to South African imports may, however, negatively affect horticulture, since South Africa probably has a strong comparative advantage in the production of most horticultural commodities arising from the capital and skill intensities of these processes and the country's long tradition of production of these commodities. Horticulture accounts for almost 40 percent of total arable production in South Africa. Competition for exports to the EC, if South Africa is included in the Lomé Convention, may also reduce opportunities for continued expansion of horticultural exports from SADCC countries. On the other hand, the South African market may be very useful for regional exporters of tea, coffee, cashew nuts, and various tropical products, and here Tanzania would benefit from the changes in the political situation that have effectively closed off South Africa as a trading partner in the past.

The output and value of South Africa's principal commodities in 1990 are shown in Table 14.1 along with the producer prices for these commodities at intervals from 1981 to 1990.

When the U.S. dollar prices are compared with those given for maize in other countries in the region (Table 14.2), it is obvious that South African farmers have been paid higher prices than farmers in the rest of the region. Since these prices have been converted at official exchange rates, they do not reflect the real position; a comparison of adjusted parity prices in the four countries would be most useful.

Role of Competition and a Liberal Trade Environment in Promoting Regional Growth

Japan and other Asian countries are particularly effective in moving capital and labor out of industries subjected to increasing competitive pressure. The Western industrial nations are less effective in doing this, and developing countries find it particularly difficult. There is some skepticism about the benefits of liberalizing in the light of an increasingly protective international environment. "If trade policy in the rest of the world is increasingly protectionist[5]. . . is this tactically an appropriate time to liberalize our own trade policies through unilateral action?" (van Zyl 1984, 55). However, it is important to retain inter-

---

[5] There has been little real change in trade and agricultural policies in the EC, for example, and throughout the world nontariff restrictions to trade have been increasing despite all the GATT negotiations and talks about reducing distortions.

Table 14.1—South African agricultural output and prices, selected harvest years

Commodity	Output, 1990		Producer Price[a] 1981	1985	1989	1990
	(US$ million)	(1,000 metric tons)	(US$/metric ton)			
Maize	948	9,442	140.42	83.22	102.96	100.70
Wheat	334	2,005	255.85	122.36	171.99	193.42
Sunflower	153	609	315.35	136.04	224.25	246.24
Tobacco (US¢/kilogram)[b]	131	32	(295.12)	(205.58)	(366.99)	(411.54)
Cotton	65	135	609.28	318.44	395.85	478.04
Dry beans	62	113	631.89	339.72	438.75	352.26
Barley	39	291	210.63	101.84	124.02	133.00
Groundnuts	34	80	528.36	262.20	332.67	381.90
Sorghum	31	368	109.48	74.86	81.12	77.90
Soybeans	25	110	349.86	149.72	209.82	222.30
Horticulture	1,499	...	...	...	...	...
Beef (US¢/kilogram)[b]	823	587	(240.38)	(86.64)	(188.37)	(180.88)
Sheep/goats	316	170	232.05	97.28	207.09	201.40
Pork	141	127	183.26	81.70	141.57	128.82
Milk (US¢/liter)[b]	361	2,430	(29.75)	(13.3)	(21.45)	(22.04)
All agriculture total	7,422	...	...	...	...	...

Source: Effective Farming, *Effective Farming Directory*, vol. 6 (Pietermaritzburg, South Africa: Effective Farming Publications, 1991).

[a] The exchange rates used are 1.19 for 1981, 0.38 for 1985 and 1990, and 0.39 for 1989.

[b] Numbers in parentheses are in U.S. cents per kilogram or per liter, as indicated next to the name of the commodity.

national competitiveness, and there is an urgent need for all southern African countries to find a means of assisting the movement of capital and labor from industries (or firms) with low competitiveness to those with strong prospects, using the market to identify products and processes in growth and in decline.

New, advanced technologies require an extensive research and development base that South Africa's small population and shortage of skills did not warrant, but for political security, heavily protected industries abound. However, with South Africa integrated into a larger SADCC region, it may be possible to develop the necessary depth. Even if there is a reluctance to liberalize unilaterally, it may make sense for an emergent South Africa to participate in the regional structural-adjustment and market-liberalization programs. The current

Table 14.2—Maize producer price for Malawi, Zambia, Zimbabwe, and South Africa, 1982-86

Year	Malawi	Zambia	Zimbabwe	South Africa	World
			(US$/metric ton)		
1982	95	142	120	203	126
1983	102	113	100	179	126
1984	87	91	100	120	130
1985	72	89	87	98	133
1986	61	128	112	117	107

Sources: Table 3.10 of this report for Malawi, Zambia, Zimbabwe, and world prices; and Effective Farming, *Effective Farming Directory*, vol. 6 (Pietermaritzburg, South Africa: Effective Farming Publications, 1991) for South African prices.

government has begun the process by reducing agricultural subsidies and reducing some of the quantitative restrictions on imports of certain agricultural commodities (for example, dry beans and beef from some countries in the region).

The question is, How will the new South African government view these issues? If South Africa is to play an effective role in regional development, trade barriers and market interventions will have to be reduced, but the underlying philosophy of the revolutionary parties favors the command economy and the need for self-sufficiency. Even if these were to be replaced with a more laissez-faire approach, it is uncertain that the government would have the confidence to allow liberal forces the freedom from regulation necessary to allow the competition so essential for efficient markets. The new government will be under enormous pressures to move toward redistribution and more egalitarian structures by state decree. The removal of regulations on marketing, trading, land size, and so forth, which have probably affected both growth and equity negatively, would in most cases also lead to greater social justice, but deregulation does not have the same political impact as direct transfers. The new government will need to be seen to be meeting the aspirations of its various ethnic groups. Deregulation also reduces the opportunity for the nepotism and corruption that is often considered a reward after years of struggle.

Although the SADCC country structural-adjustment programs emphasize the reduction of all tariffs (for example, Zimbabwe is supposed to reduce all tariffs to between 0 and 30 percent by 1994), it is uncertain whether these tariff reductions will, or indeed should be, achieved. Given the very high income disparities that exist in southern Africa, opening access to foreign currency would result in very strong demand for luxury commodities not produced in Zimbabwe or even in South Africa.

There is a very high marginal propensity to import luxury commodities in the region—particularly where the main initial beneficiaries of devaluation (the first step in the Economic Structural Adjustment Program) are in the highest income brackets.[6] Very high rents are currently earned on luxury commodities. It would seem sensible for the government to place very high tariffs on these commodities, earn the rents, and reduce the demand for foreign currency. This is one of the major constraints faced by countries in the region that want to free access to foreign currency. Restricting demand for luxuries would assist in moving away from controls and licenses for foreign currency and encourage more expenditure on production inputs. Tariffs on goods such as videos, microwaves, and hi-fi equipment in Zimbabwe would need to be over 1,000 percent if they were to have any effect on reducing demand.

To the extent that tariffs are not reduced on products that are, or could be, produced within the SADCC region, the regional "pie" is reduced. However, a much slower phasing-out of tariff barriers on luxury goods would seem a small sacrifice if it enabled governments to reduce, with confidence, the controls and licenses that have stifled investment and development in the region for so long.

While it is true that successful macroeconomic policy helps to make sectoral competitive problems easier, relatively low factor productivity and lack of competitiveness are among the underlying microeconomic causes of the macroeconomic malaise. The main problem with most of the southern African economies is that they do

---

[6] The devaluation of the Zimbabwean dollar since March 1991 has resulted in an increase of around Z$1 billion to be distributed among some 1,500 farmers—assuming that the full effect of the devaluation has been passed on to producer prices. Tax revenues will increase, and it is possible that some of the farm labor will benefit where bonuses are paid, but the largest proportion of the revenues will increase the demand for urban property and luxury consumer goods. This is particularly true in an environment that does not encourage investment because of the shortage of materials for on-farm investment, barriers to entry in industry, and the anticipation of continued high inflation.

not produce enough goods and services to meet demand. They are not addressing the problem of relative scarcity.

In South Africa the ratio of capital to labor has increased steadily, but capital productivity has declined since 1970, thus reducing both employment and growth (Truu 1986). Truu goes on to point out that in South Africa this reduced efficiency has not been balanced by more equity, since the economy "has not been so much concerned with distributing the desirable things in scarce supply as with actually reducing their availability." While this statement was made about the effect of sanctions, it is equally true of all the countries in the region, including South Africa, that the heavily regulated economies have in fact resulted in inefficiencies that have reduced growth, equity, and employment.

CONCLUSION

The southern African economies have been operating so far within the welfare frontier that many measures that could be taken would not necessarily result in the classic efficiency/equity trade-off.[7] A reduction of the large rents accruing to those with access to foreign currency, licenses, and so forth, in the region would go a long way toward increasing growth and equity. The progress in Zimbabwe's structural-adjustment program, while severely restricting the supply of money, is lagging seriously behind in any measures to reduce government spending or to reduce rents by deregulation. Unless government deficits are reduced, the only means of reducing inflation is to severely restrict the money supply, resulting in very high interest rates, which in turn reduce investment. Morande and Schmidt-Hebbel (1991) show that there is a significant correlation between fiscal policy and private investment in Zimbabwe.

Unless the barriers to competition can be removed, the recent devaluations will not achieve the desired impact on growth and could simply act as a fuel for inflation. The devaluation will encourage exports and import-substitution, but the multiplier effects that are such an important part of the strategy will not be achieved until the barriers to entry have been removed and access to inputs (foreign currency), competition, and the investment environment improved.

---

[7] An example is given in Muir-Leresche 1984 of changes to crop price and wage policy in Zimbabwe that would increase equity, employment, and efficiency.

The inflationary impact of devaluation depends on institutional factors and is particularly likely in monopoly situations with barriers to entry. Regarding employment, agricultural exports tend to have higher labor intensities than the nontraded commodities, and to the extent that the devaluation encourages these exports there will be positive effects on employment. As unskilled labor has a high propensity to consume locally produced goods if the supply constraints have been lifted, this could have very strong multiplier effects throughout the economy. The problem arises when the supply constraints have not been lifted and the increased employment simply results in higher inflation. These problems must be addressed or the programs will fail entirely, and the devaluations could result in an uncontrolled devaluation/inflation spiral. The devaluations are certainly an essential condition for growth, but they are not a sufficient condition.

Most of the countries of the region, including South Africa, have very distorted economies with prices rarely reflecting opportunity costs. Until domestic markets are freed, barriers to trade are lifted, and the region becomes more attractive to investors, a politically acceptable South Africa is unlikely to have much impact on agricultural trade and development within the region.

REFERENCES

Effective Farming. 1991. *Effective farming directory*, vol. 6. Pietermaritzburg, South Africa: Effective Farming Publications.

Jansen, D. 1988. *Trade, exchange rate and agricultural pricing policies in Zambia*. A World Bank Comparative Study in The Political Economy of Agricultural Pricing Policy series. Washington D.C.: World Bank.

Mill, J. S. [1859] 1937. On liberty. Quoted in *A History of Political Theory*, ed. G. H. Sabine, London: Harrap and Co.

Morande, F., and K. Schmidt-Hebbel. 1991. *Macroeconomics of Public Sector Deficits: The Case of Zimbabwe*. WPS 688. Washington D.C.: World Bank.

Muir-Leresche, K. 1984. Crop price and wage policy in the light of Zimbabwe's development goals. Ph.D. thesis, University of Zimbabwe, Harare.

Muir-Leresche, K., and D. Jansen. 1991. Trade, exchange rate policy, and agriculture. Conference on Zimbabwe's Agricultural Revolution, July, Victoria Falls, Zimbabwe.

Truu, K. L. 1986. Economics and politics in South Africa today. *The South African Journal of Economics* 54 (4).

van Zyl, J. C. 1984. South Africa in world trade. *The South African Journal of Economics* 52 (1).